DRIVERS IN THE 1980s

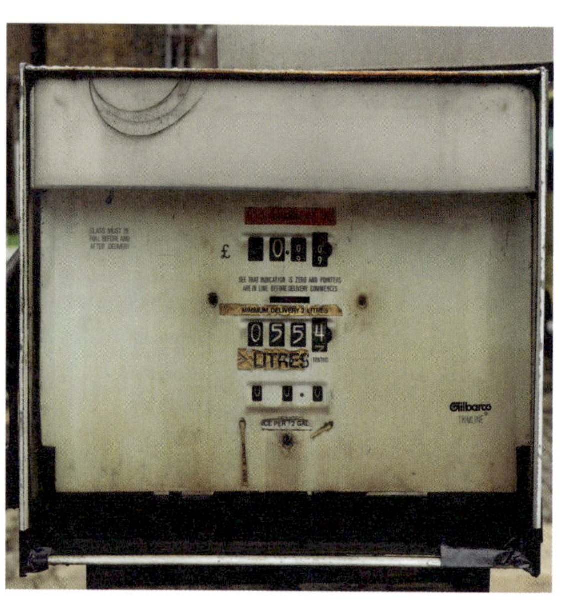
£
SEE THAT INDICATION IS ZERO AND POINTERS
ARE IN LINE BEFORE DELIVERY COMMENCES
0554
LITRES
Gilbarco

DRIVERS IN THE 1980s

by Chris Dorley-Brown

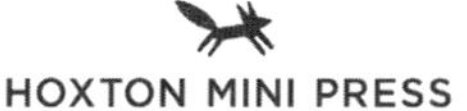

HOXTON MINI PRESS

Book One

I've Lived in East London for 86½ Years

Book Two

East London Swimmers

Book Three

A Portrait of Hackney

Book Four

Shoreditch Wild Life

Book Five

One Day Young in Hackney

Book Six

Drivers in the 1980s

East London Photo Stories

Book Six

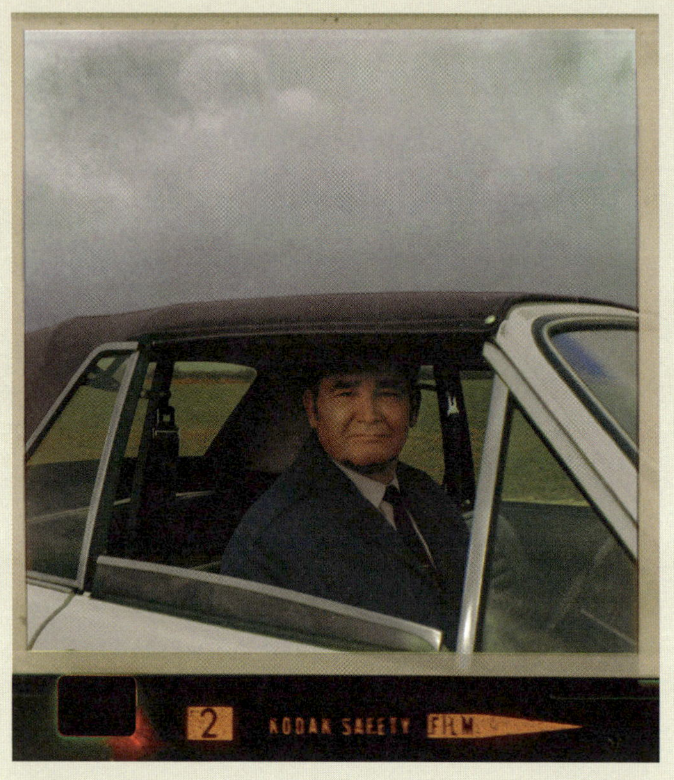

Peter Dorley-Brown, Ford Cortina MkII, 1967

INTRODUCTION

I grew up around cars. Dad was a car man; he bought them, sold them, mended and polished them and sold the petrol to make them run. On Saturday nights he drove a breakdown truck; the police had him on a rota and when the boozers closed at 11pm sharp the predictable carnage would begin. He would follow the ambulance and fire crews and pick up the bits of glass, rubber, metal, sometimes even the bits of people. He towed the motors back to the yard where, a few days later, as a schoolboy, I would prise open the deformed doors and sit in the drivers' seats and imagine the worst. Broken safety glass under my backside, smears of blood on the vinyl, clocks stopped at the moment of impact, an eerie prelude to a novel I was shortly to read which I imagined had been written especially for me.

When I was eight years old, dad arrived home one day in the most glamorous Ford I had ever seen: a convertible Cortina Mk2, off-white with a black canvas roof. We piled in and went for a spin on country roads and at one point stopped to admire the view. I picked up mum's camera and for the first time in my life I took a photograph. It was of the old man, still in his work clothes, dark suit and tie, looking out of the car window.

By the time I made the pictures in this book, on a May morning in 1987, photography had progressed from hobby to profession via irrational obsession (it's now settled into a mixture of all three). I had long since forgotten my motivation for stepping out that day. Photographic missions with no specific agenda were standard for me and I'd never heard the word 'psycho-geography'. But the final image on the contact sheets, now re-studied and scanned after all these years for this book, reveals that my destination was to record the sell-off of Rolls Royce, one of the infamous Tory public company jumble sales going on in the City. Maybe there was a picture or two to be had that would encapsulate this event that was dominating the news.

The sell-off was one in a series of unedifying gold-rush stampedes and as I left the studio on foot and headed towards Bank

Station, the traffic on Mare Street and Hackney Road was gridlocked by people trying either to get to the sell-off, or mainly, trying to avoid it. It was the first warm day of the year and the drivers and passengers were trapped: impatient, sweaty, wanting to be outside in the sunshine. As far as I and the Rolleiflex were concerned, they were asking for it. I could get close, very close. They were sitting ducks.

I followed the traffic and walked south onto Shoreditch High Street, then all the way down to the river by London Bridge. Tourists were stuck on open top Routemaster buses, motorcycle couriers were lighting fags, engines idled, the lights would only let a few through on green before the waiting continued. The sell-off was causing chaos, drivers looked deflated, maps were consulted for escape routes, midday editions of the *Evening Standard* given the once-over, frustrated children plonked on drivers laps. The traffic islands, surrounded by those now long-gone steel barriers, provided a refuge from which to take photos, and I was more or less invisible, able to fully concentrate on these prisoners of glass, metal and capitalist reverie.

These images are from the first five rolls of colour film I ever shot. Simple portraits, faces and torsos reframed by the windscreens and doors of the vehicles, they hold a stillness in a world of movement and unpredictability.

I made a few prints, I liked one in particular of a man who I imagined was a disenchanted stockbroker sitting alone on a bus staring into a void as he absorbed the reality of another bad day at the office. He looked like a man from a bygone era, the bus windows grimy with bird shit, game over. I put the negs away, an unfinished exercise which I would revisit a year later, with another foray into the city during summer, another five rolls. Then, back in the drawer they went.

Here are all of the images, more or less. I never bought another roll of black and white film again, the journey had started. Oh, the book I mentioned earlier, it was written by J.G. Ballard: *Crash*.

Chris Dorley-Brown
Hackney, 2014

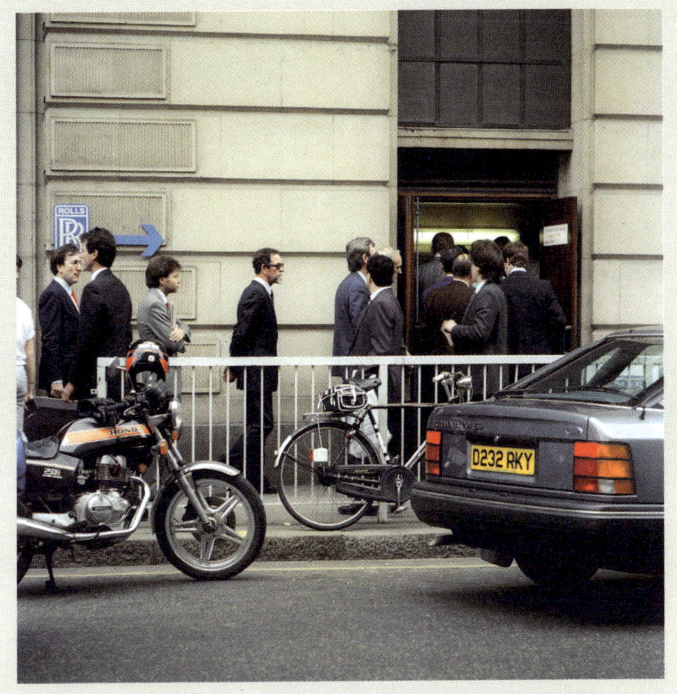

Rolls-Royce privatisation, Threadneedle Street, London, 1987

BMW 3 series, Old Street, EC1

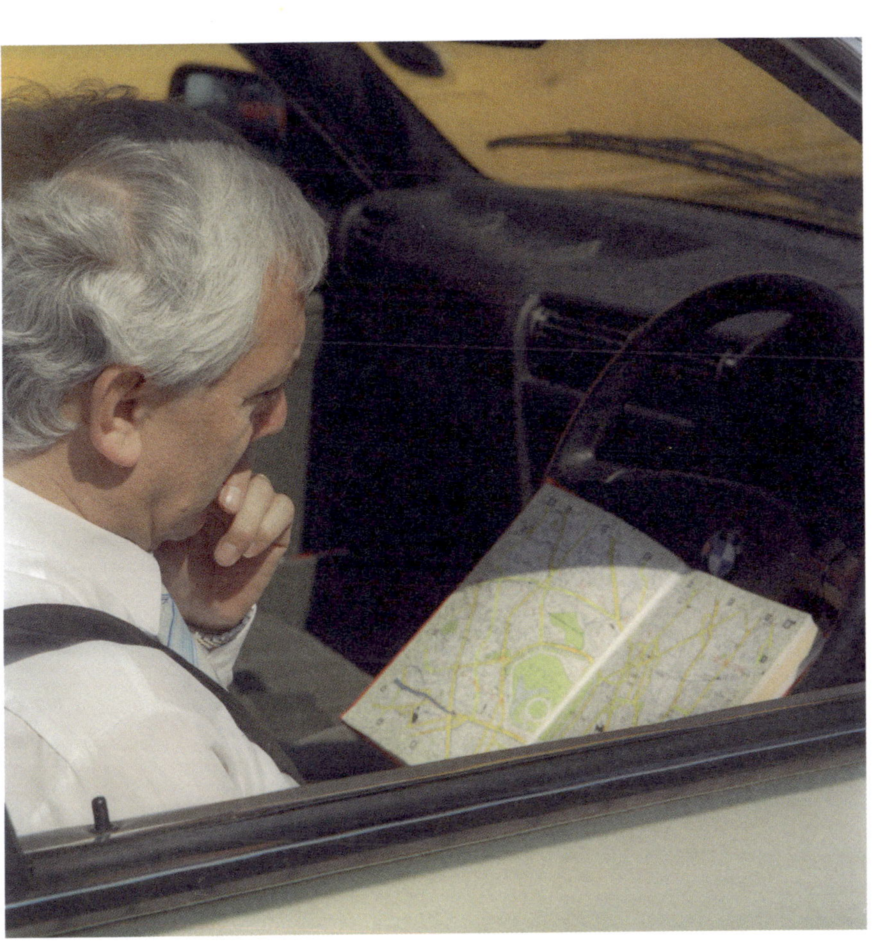

Vauxhall Astra Mk1, Piccadilly, W1

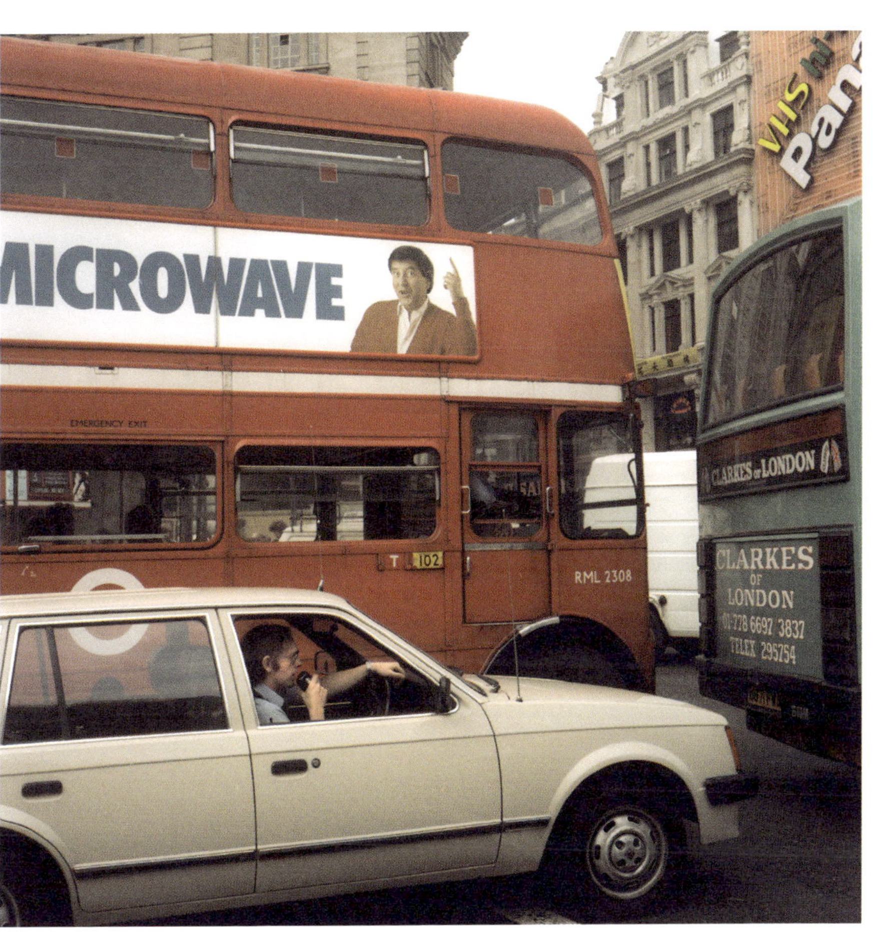
MICROWAVE
EMERGENCY EXIT
RML 2308
CLARKES OF LONDON
CLARKES
OF
LONDON
01-778 6697 3837
TELEX 295754
VHS

Austin Maestro City Van, London Bridge, EC4

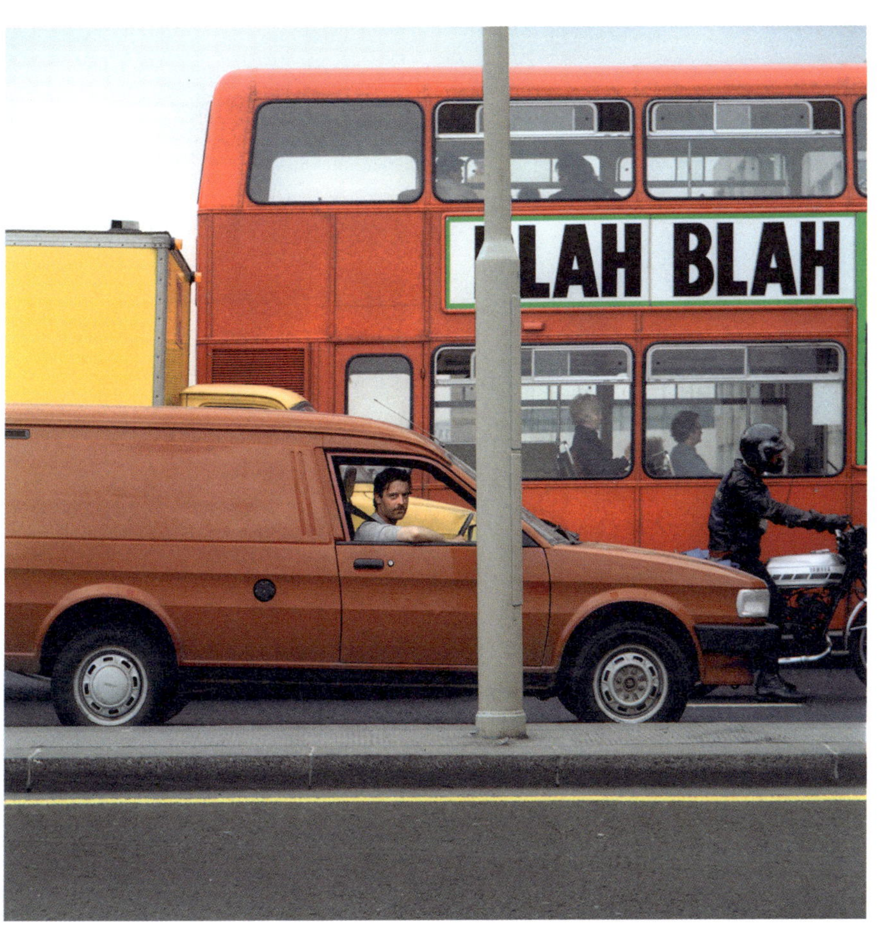
LAH BLAH

Ford Transit Mk2, London Bridge, EC4

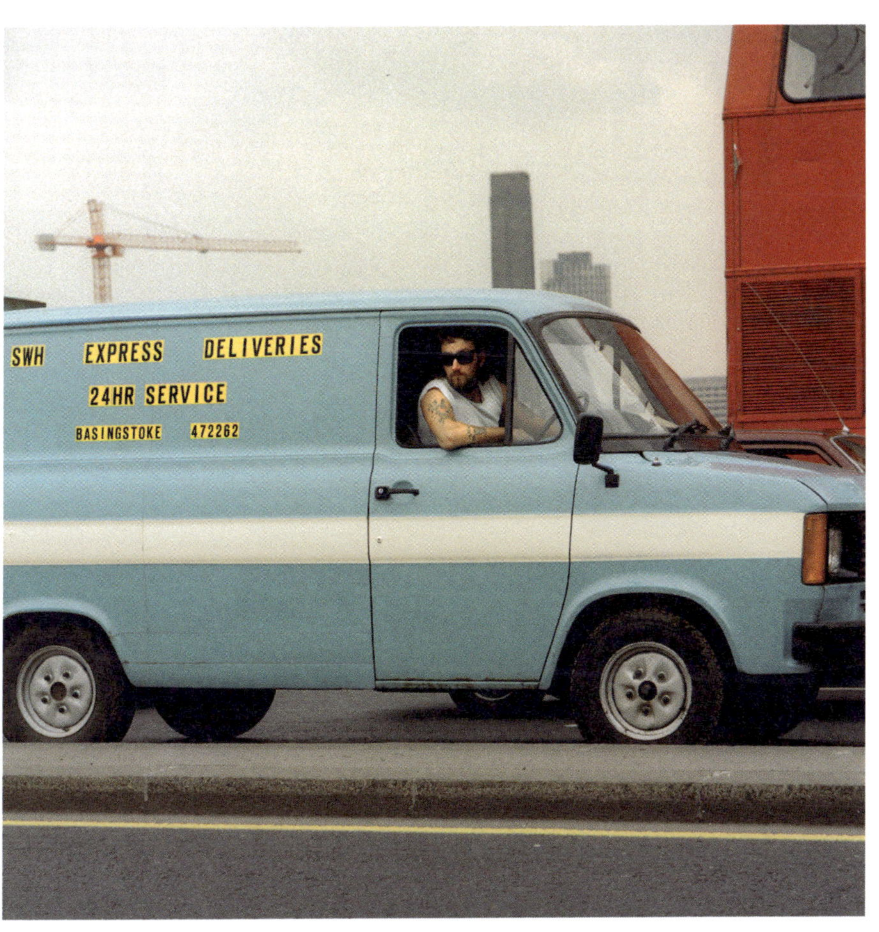
SWH EXPRESS DELIVERIES
24HR SERVICE
BASINGSTOKE 472262

Ford D-Series Tipper, Mare Street, E8

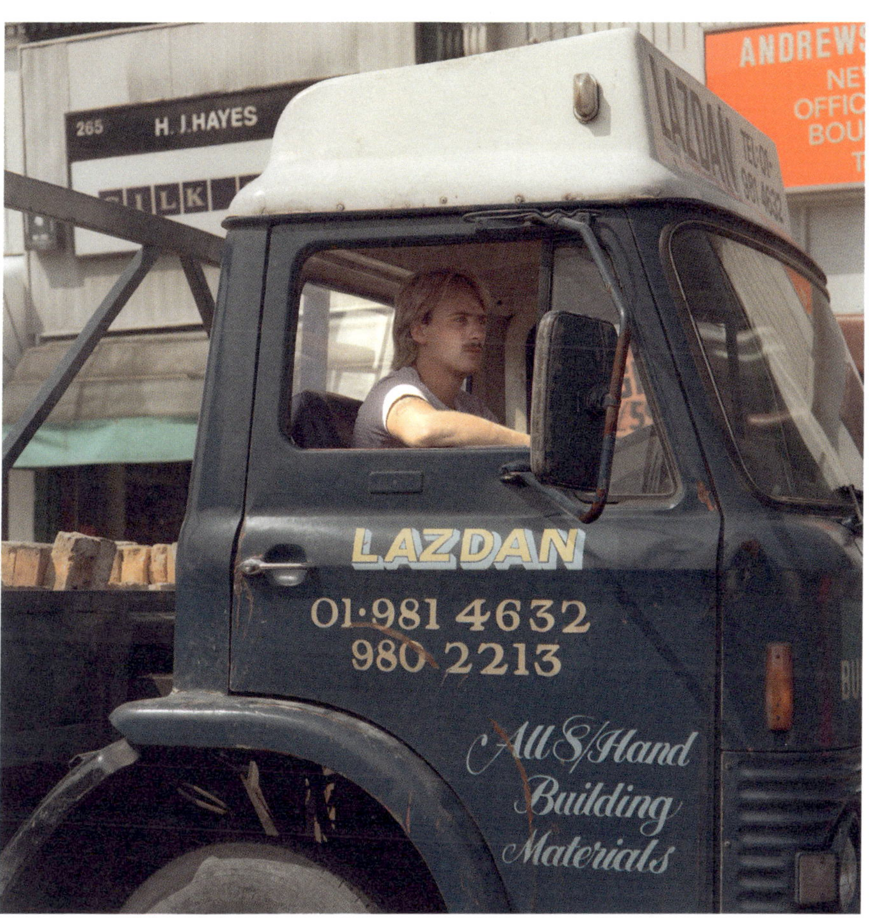
265
H.J.HAYES
ANDREWS
LAZDAN
LAZDAN
01·981 4632
980 2213
All S/Hand
Building
Materials

Leyland Sherpa, Mare Street, E8

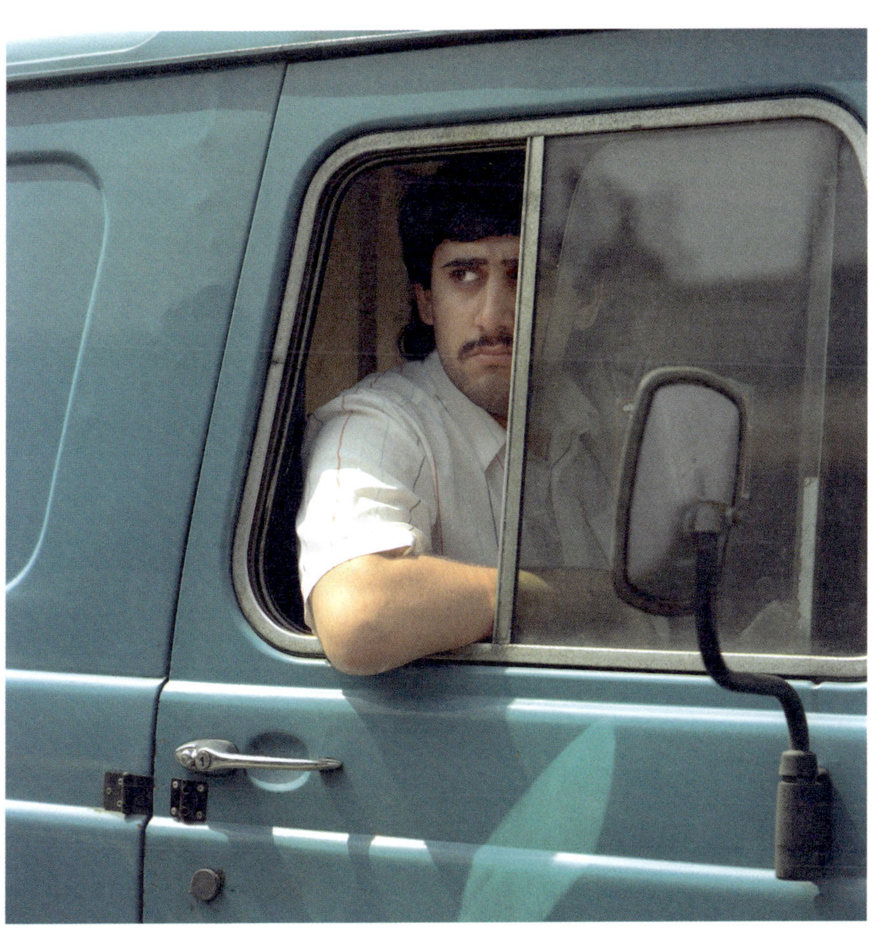

Ford Escort Mk3 Van, Hackney Road, E2

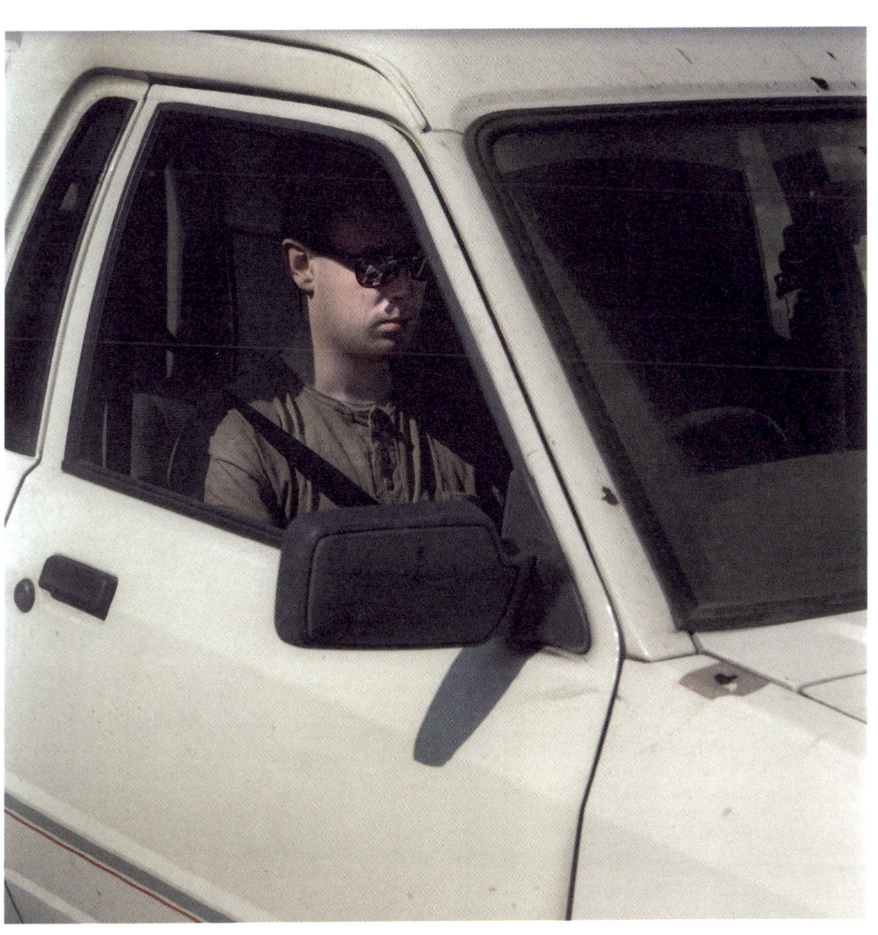

AEC RM Bus, Bank, EC2

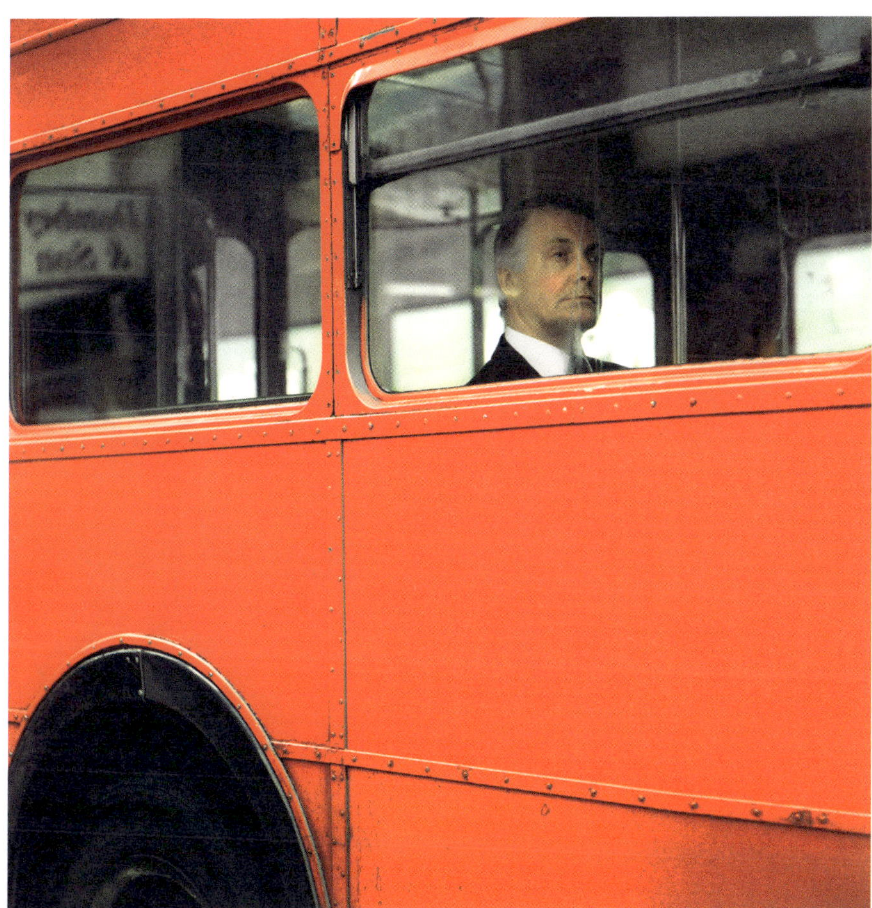

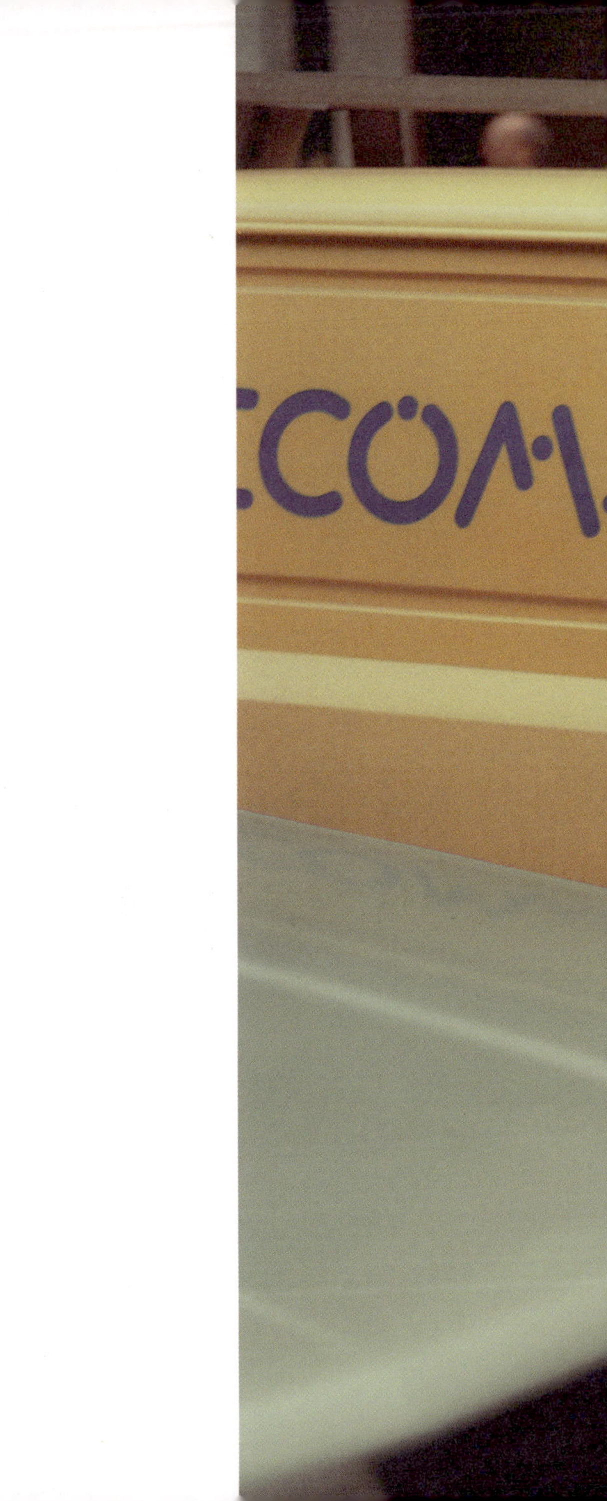

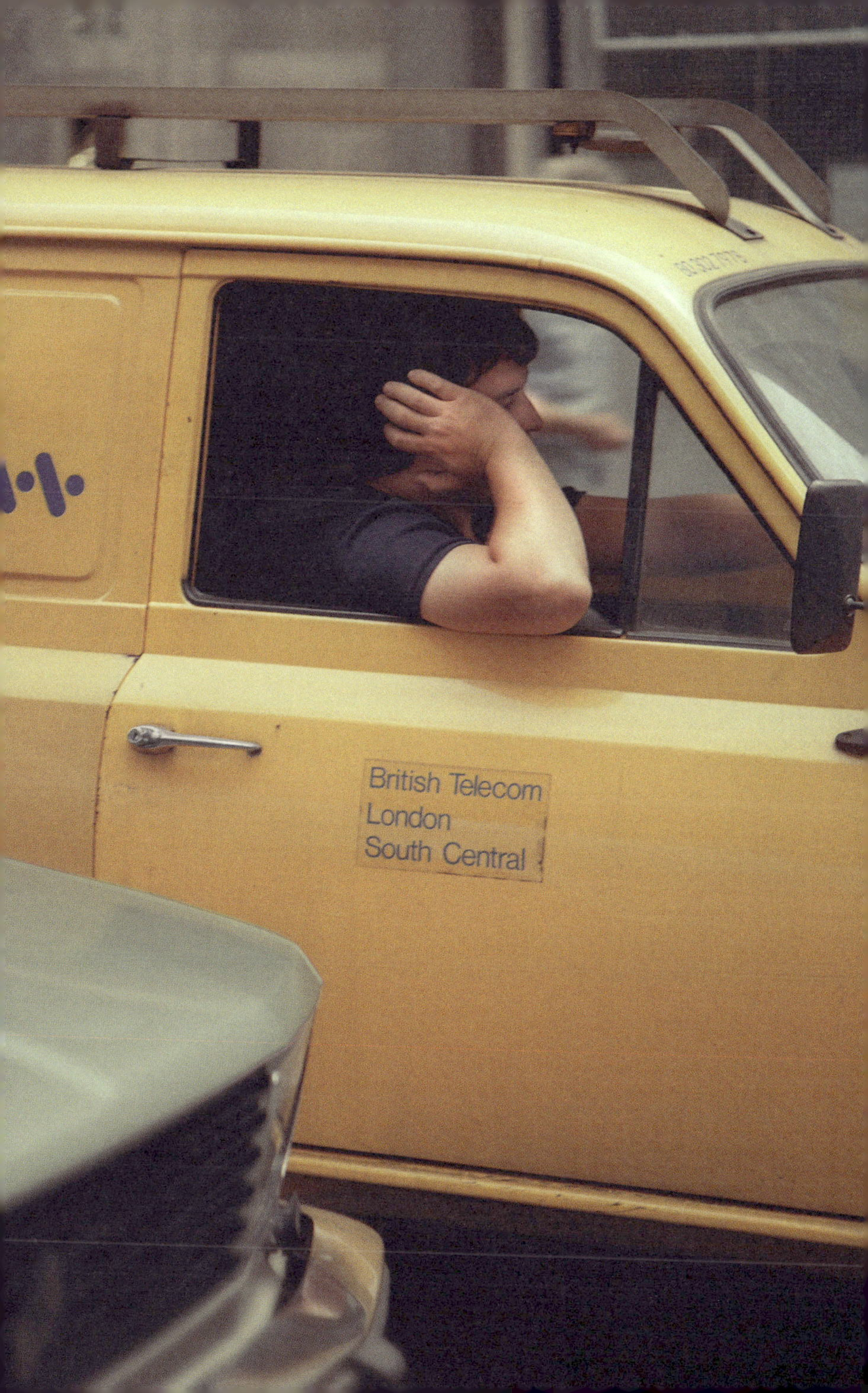
British Telecom
London
South Central

Peugeot 604, Mare Street, E8

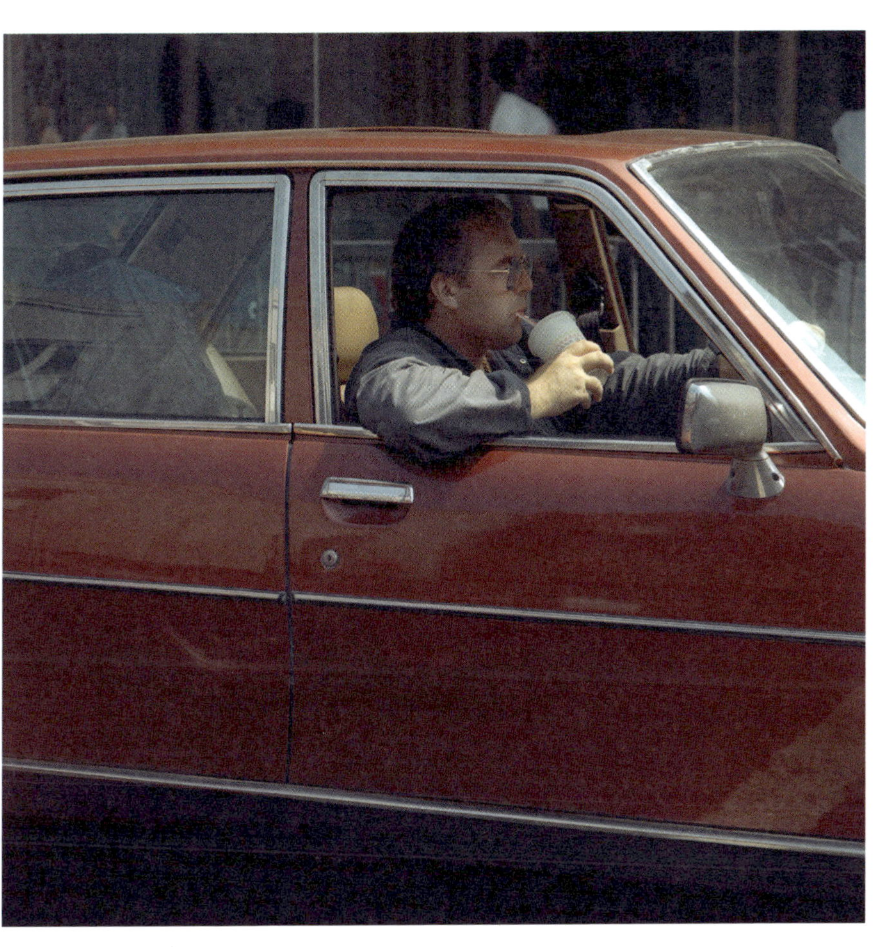

Vauxhall 560 Astramax L Van, Moorgate, EC2

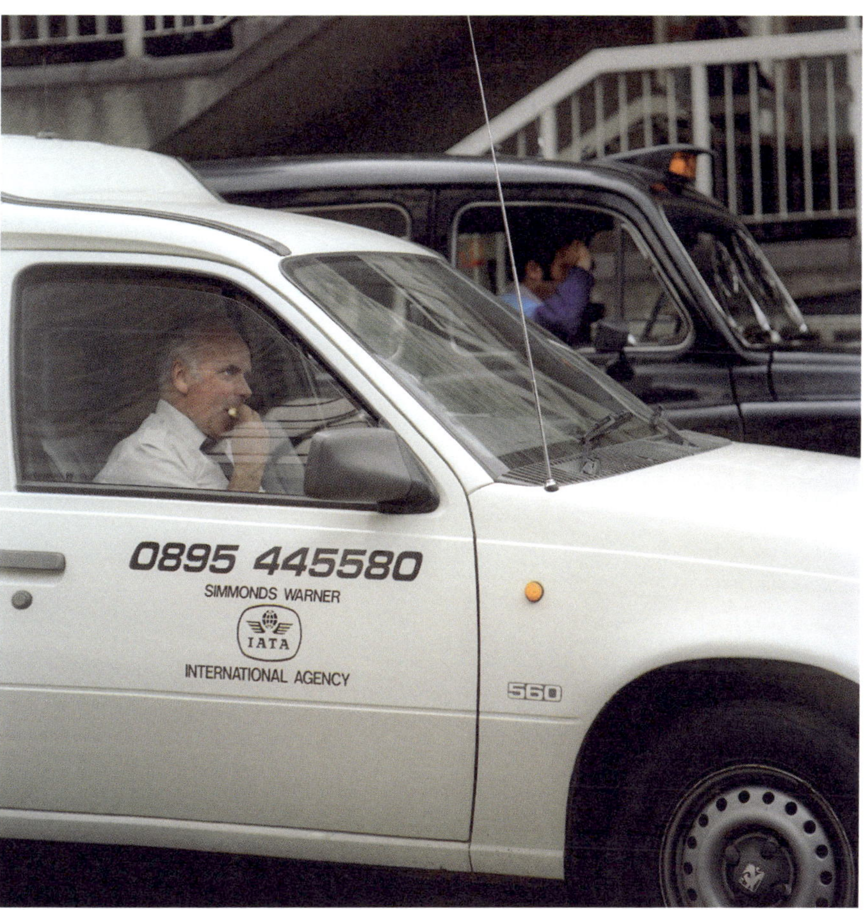

0895 445580
SIMMONDS WARNER
IATA
INTERNATIONAL AGENCY
560

MG Metro, Norton Folgate, EC2

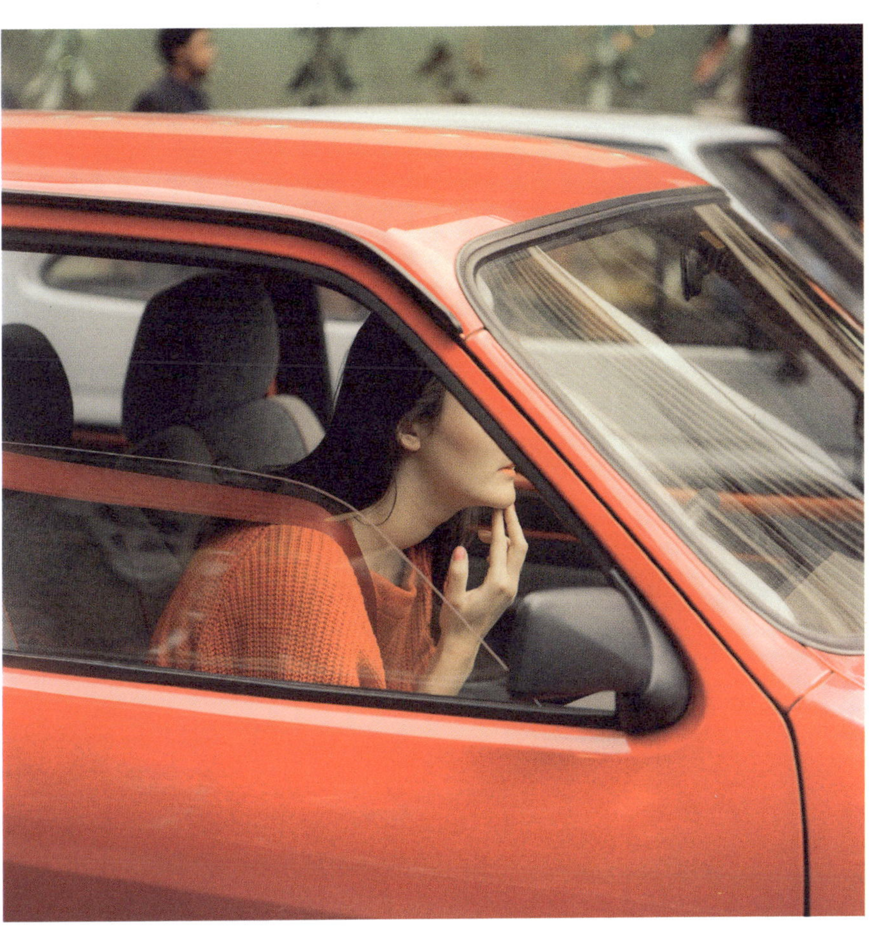

Bedford HA Van, Hackney Road, E2

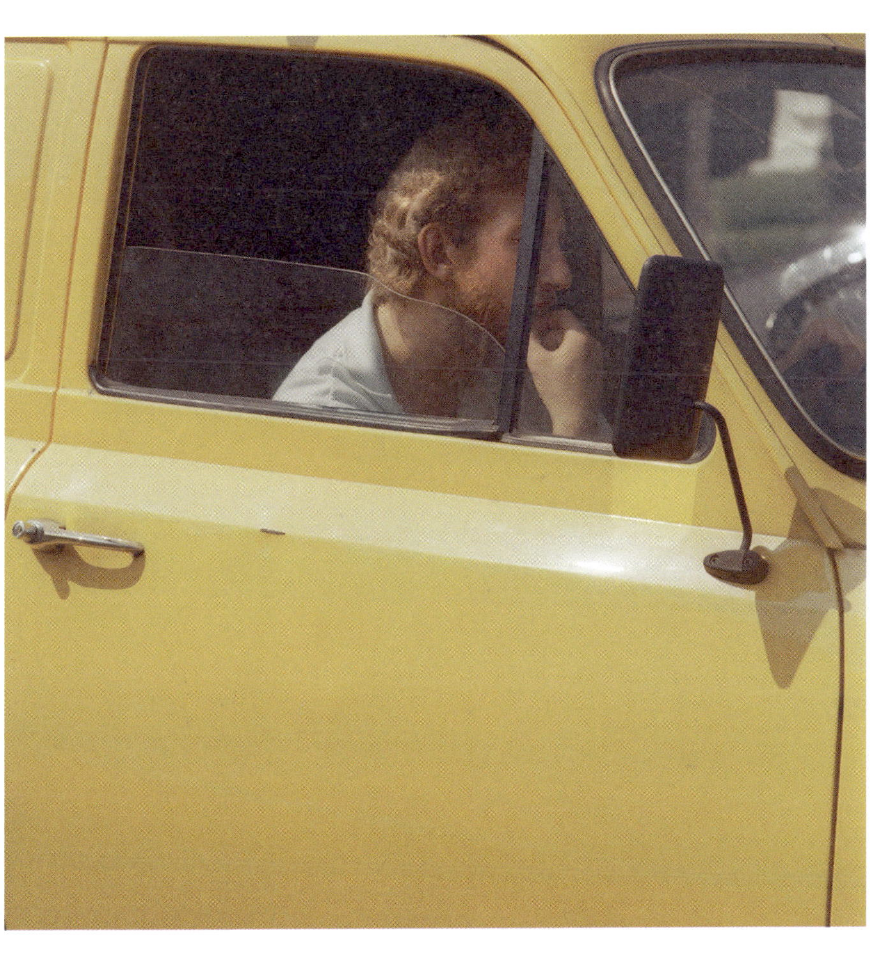

Vehicle unknown, London Bridge, EC4

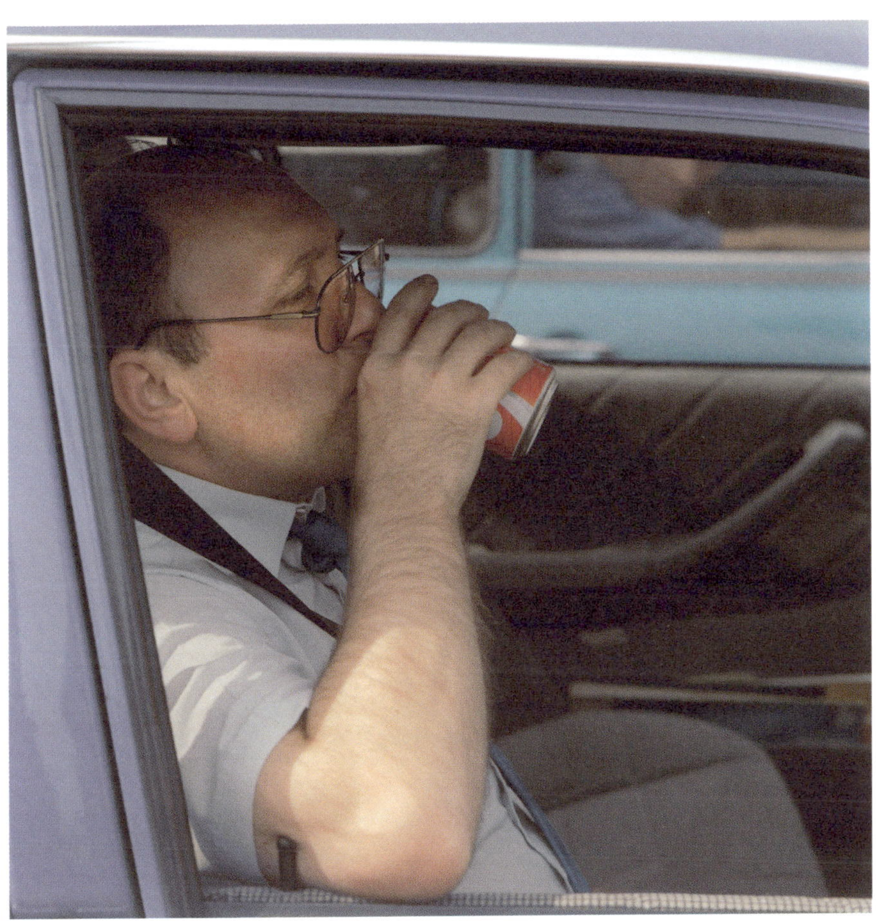

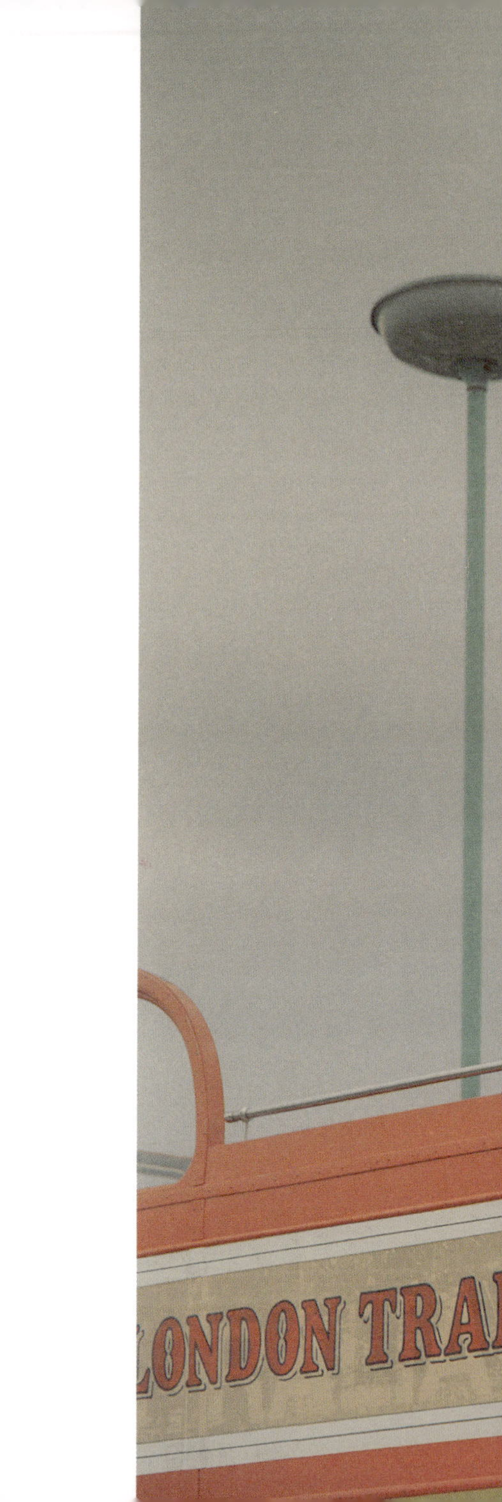
LONDON

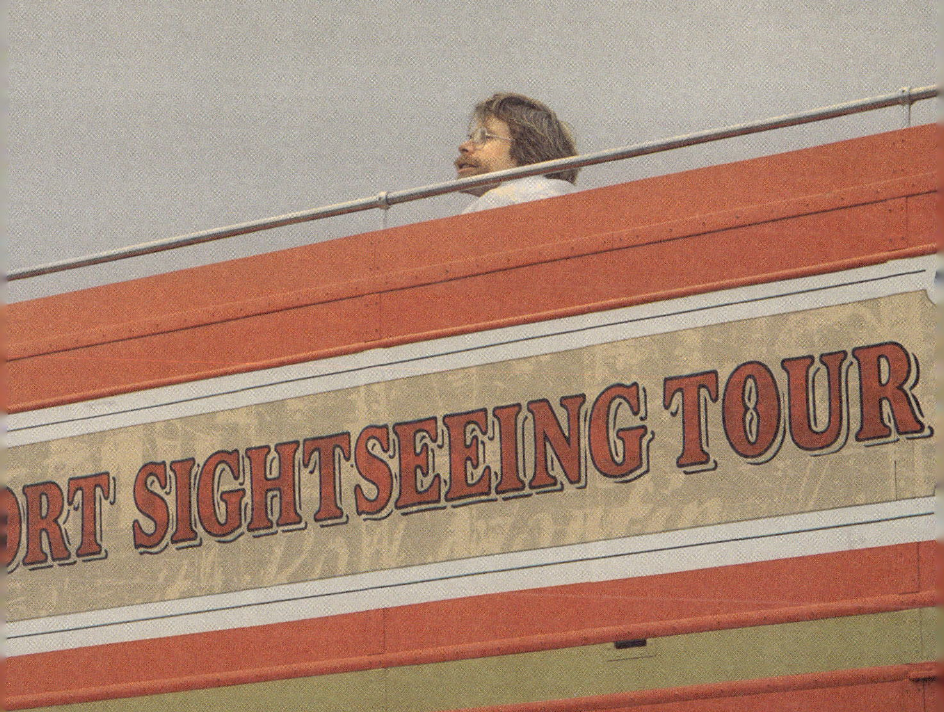
ORT SIGHTSEEING TOUR

Vehicle unknown, Bishopsgate, EC2

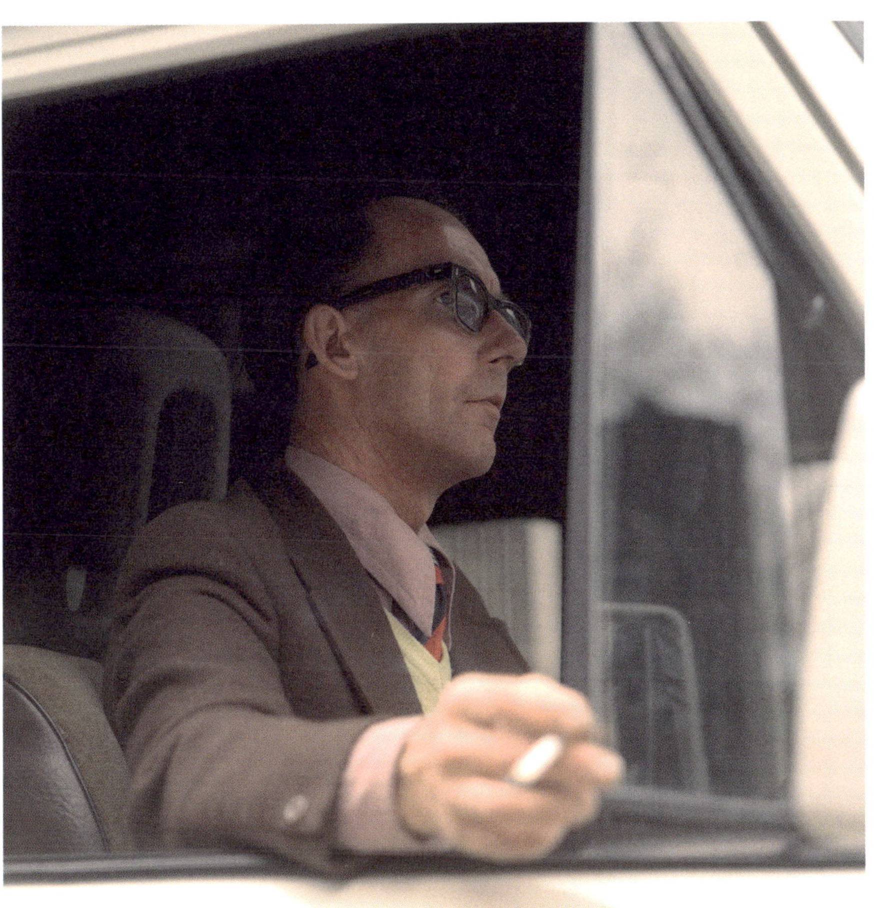

Vehicle unknown, Bishopsgate, EC2

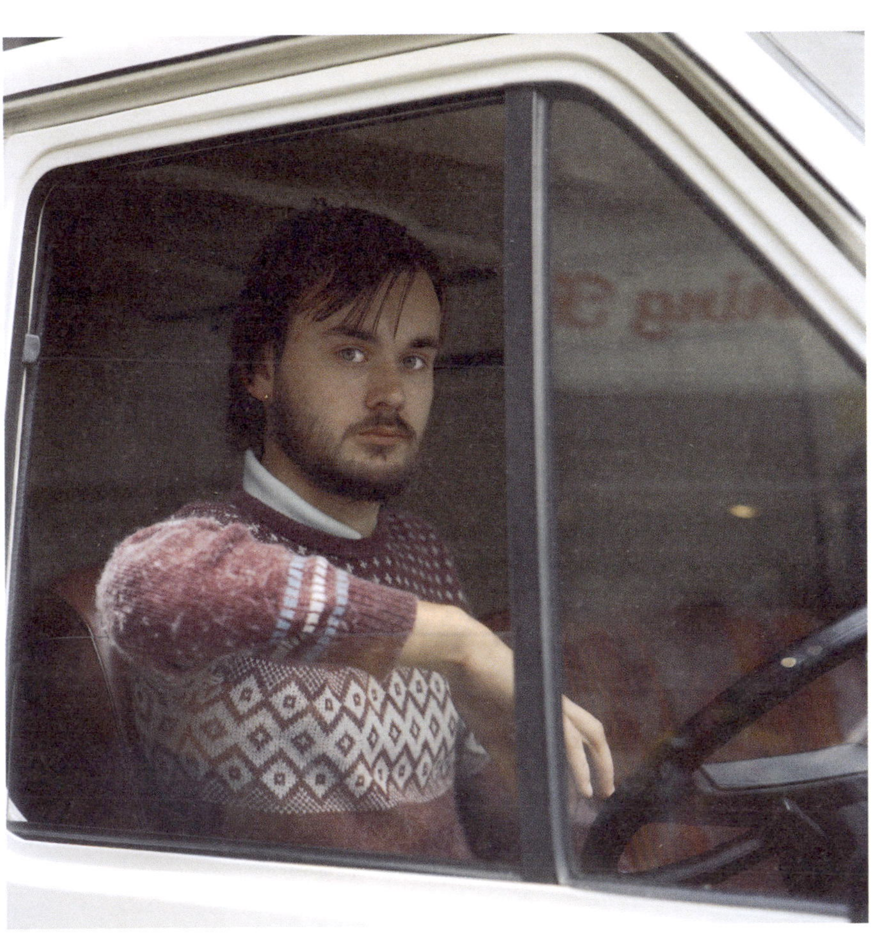

Toyota Hiace, Hackney Road, E2

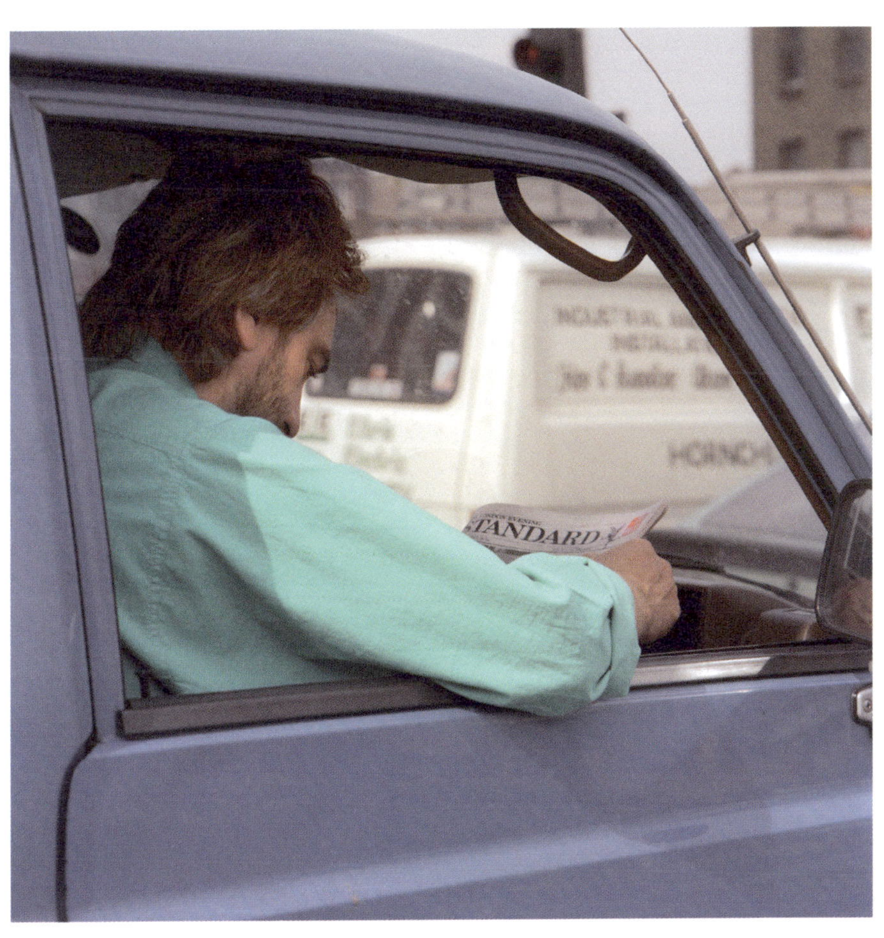
STANDARD

Ford Escort Mk3 Van, Gracechurch Street, EC3

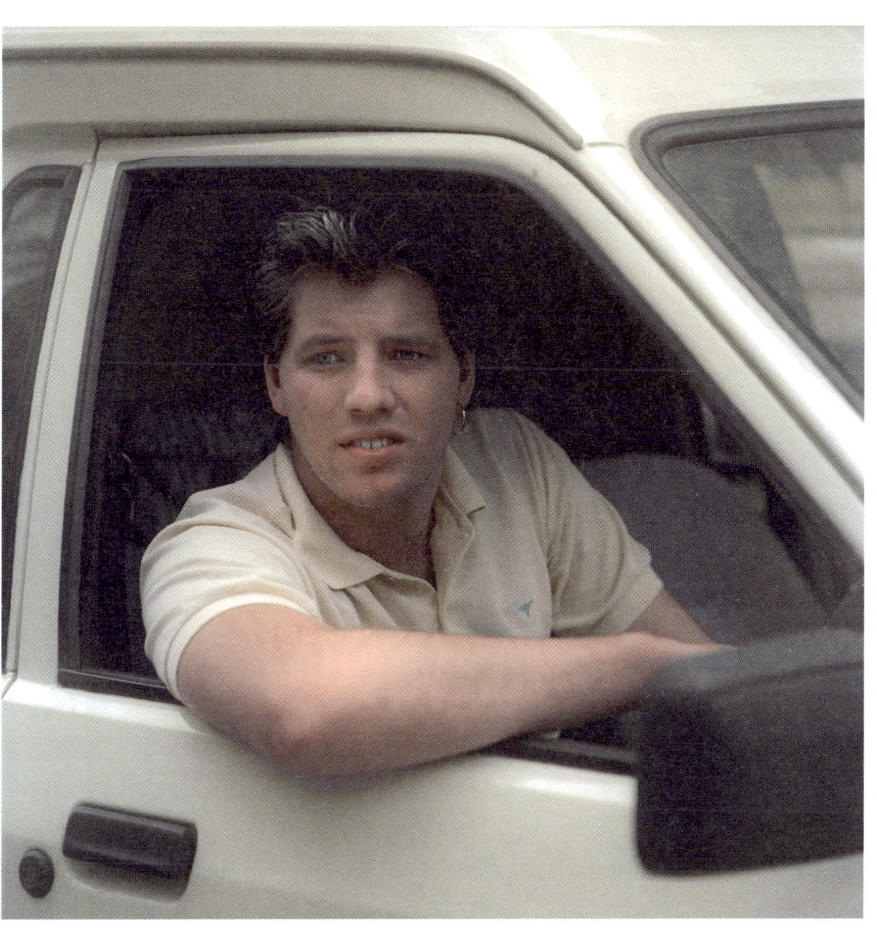

Black Cab LTI FX4, Mare Street, E8

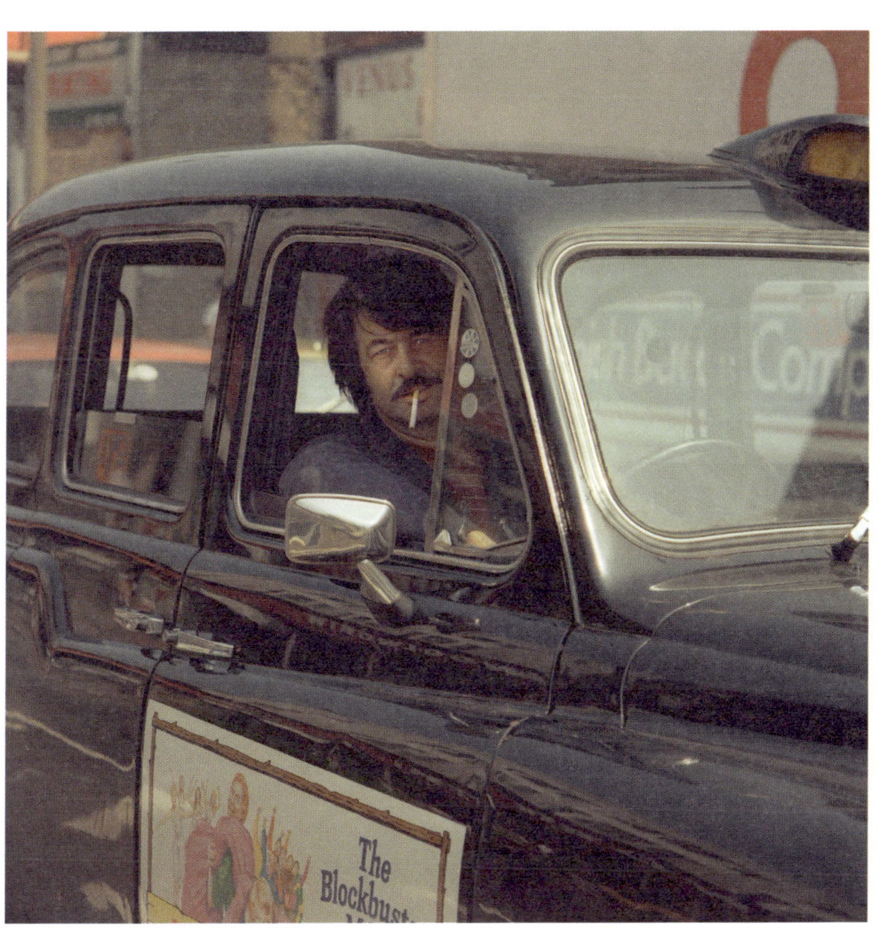
The
Blockbust

Cortina Mk4, Old Street, EC2

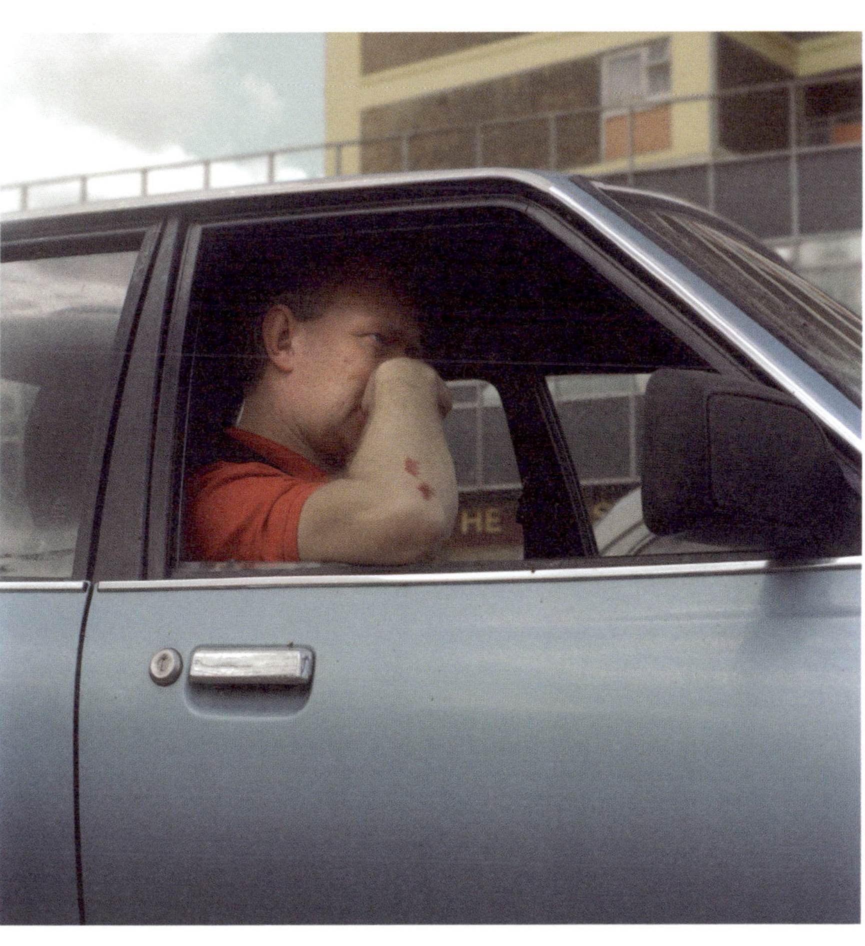

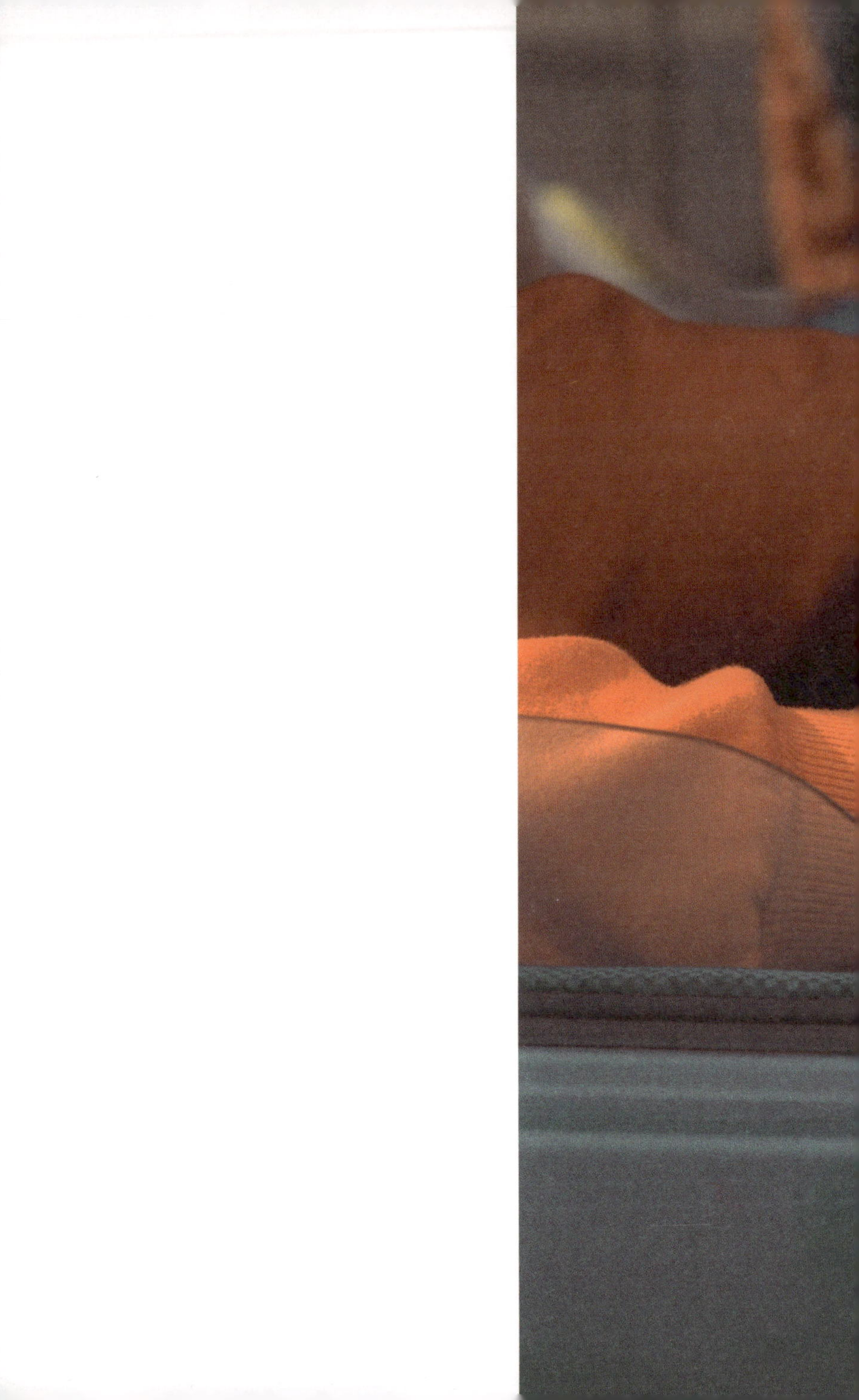

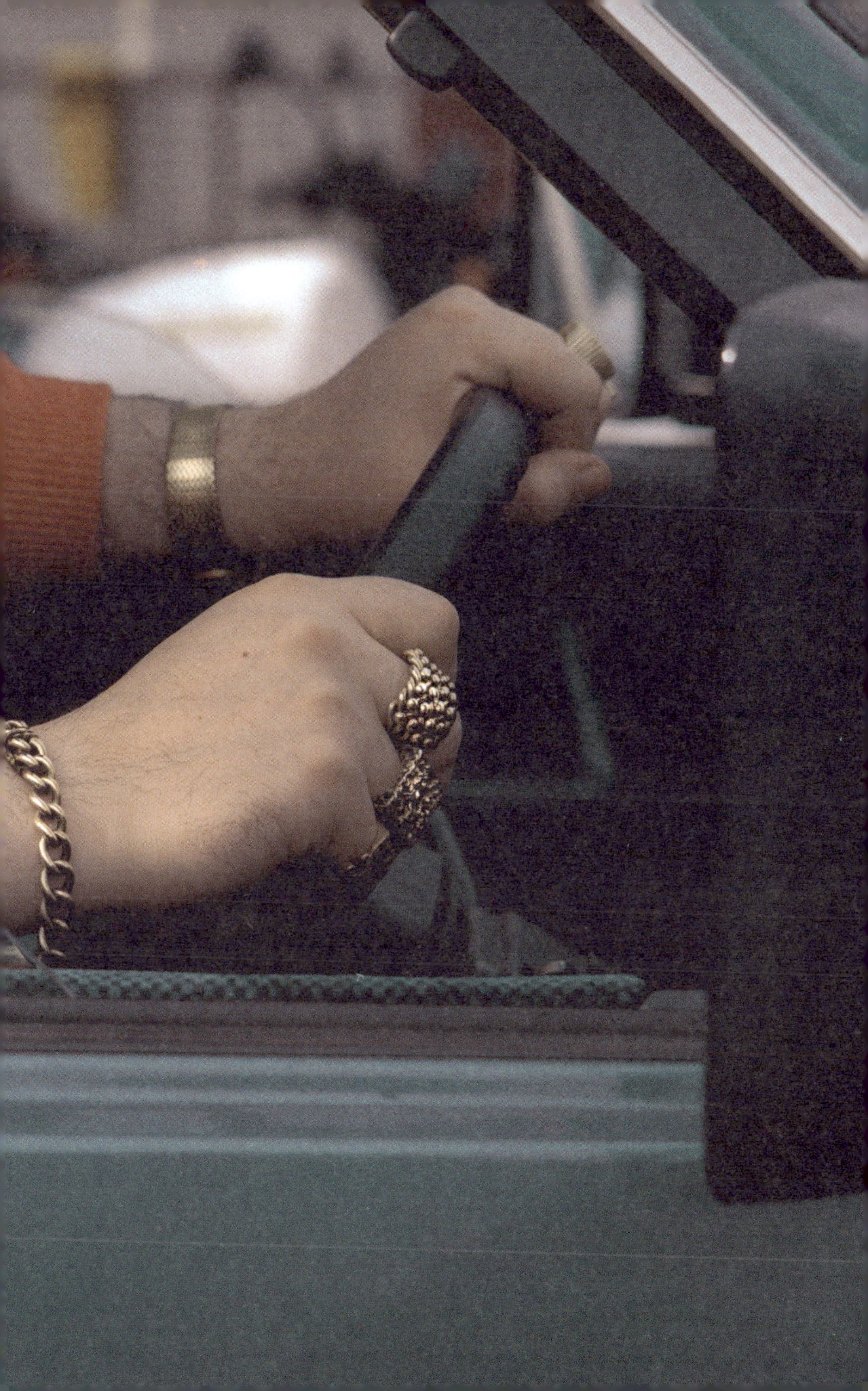

Honda 90 Moped, Mare Street, E8

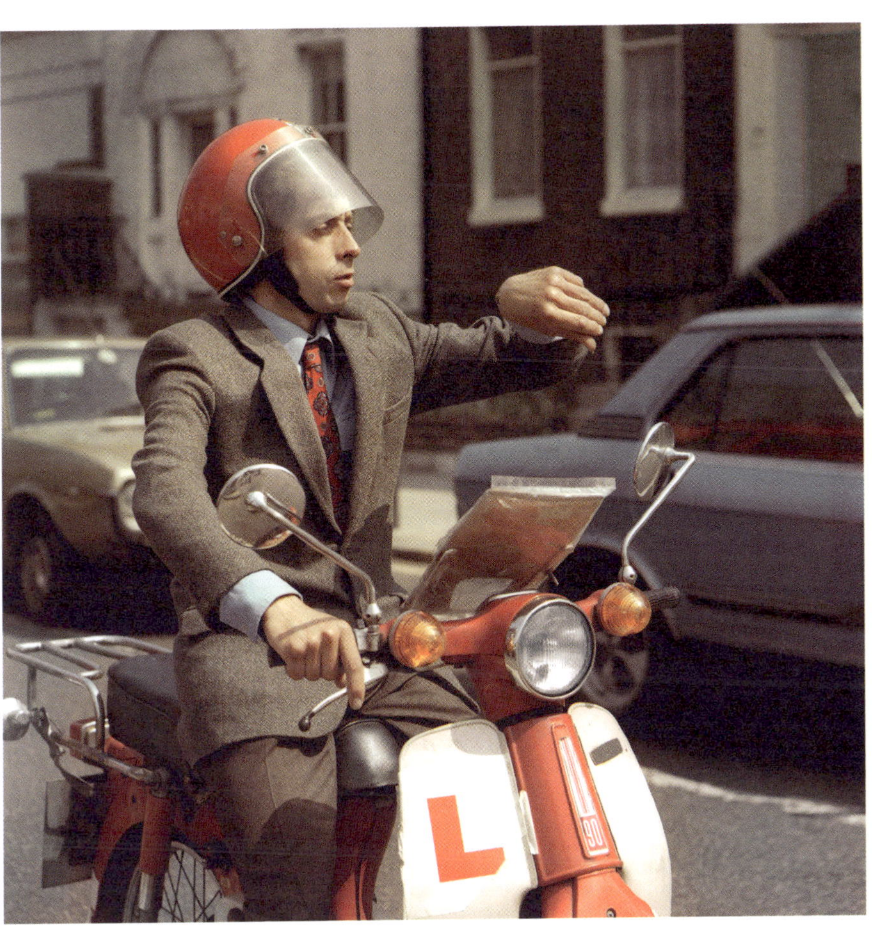
L
90

Vehicle unknown, Gracechurch Street, EC3

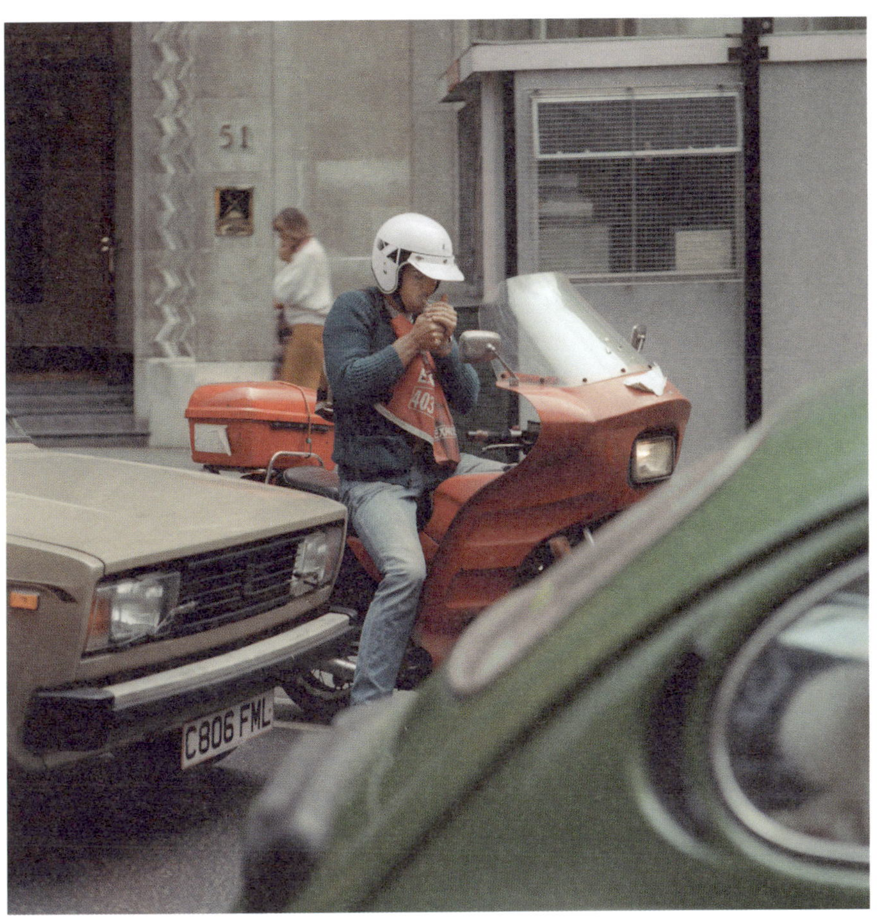
51
C806 FML

Vehicle unknown, Hackney Road, E2

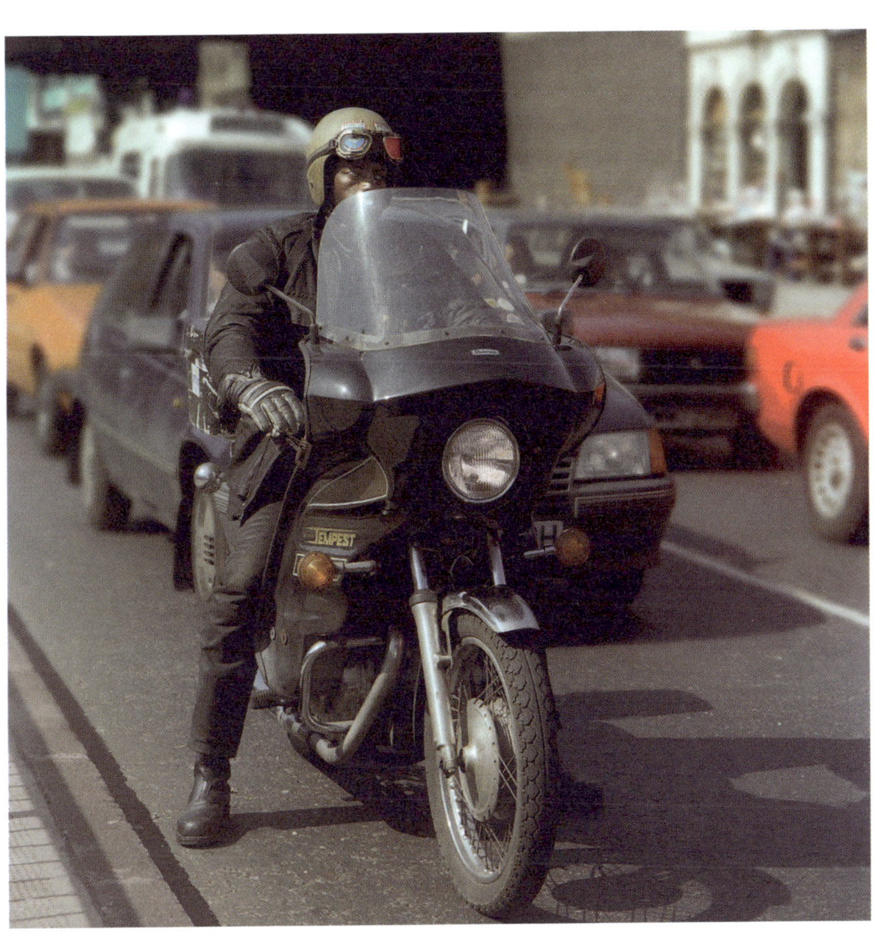
TEMPEST

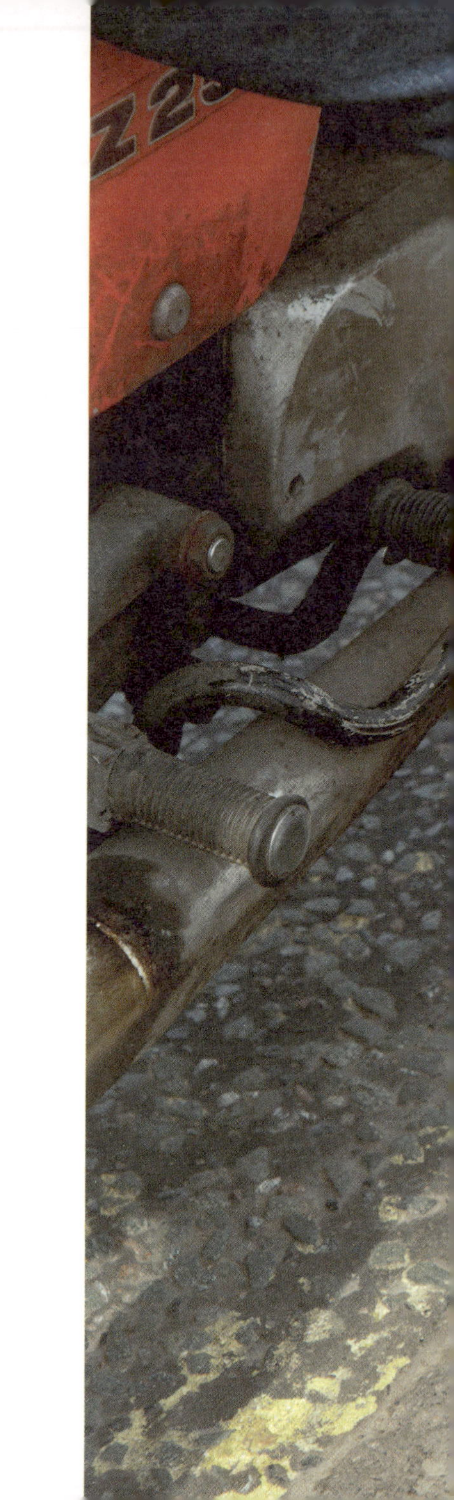

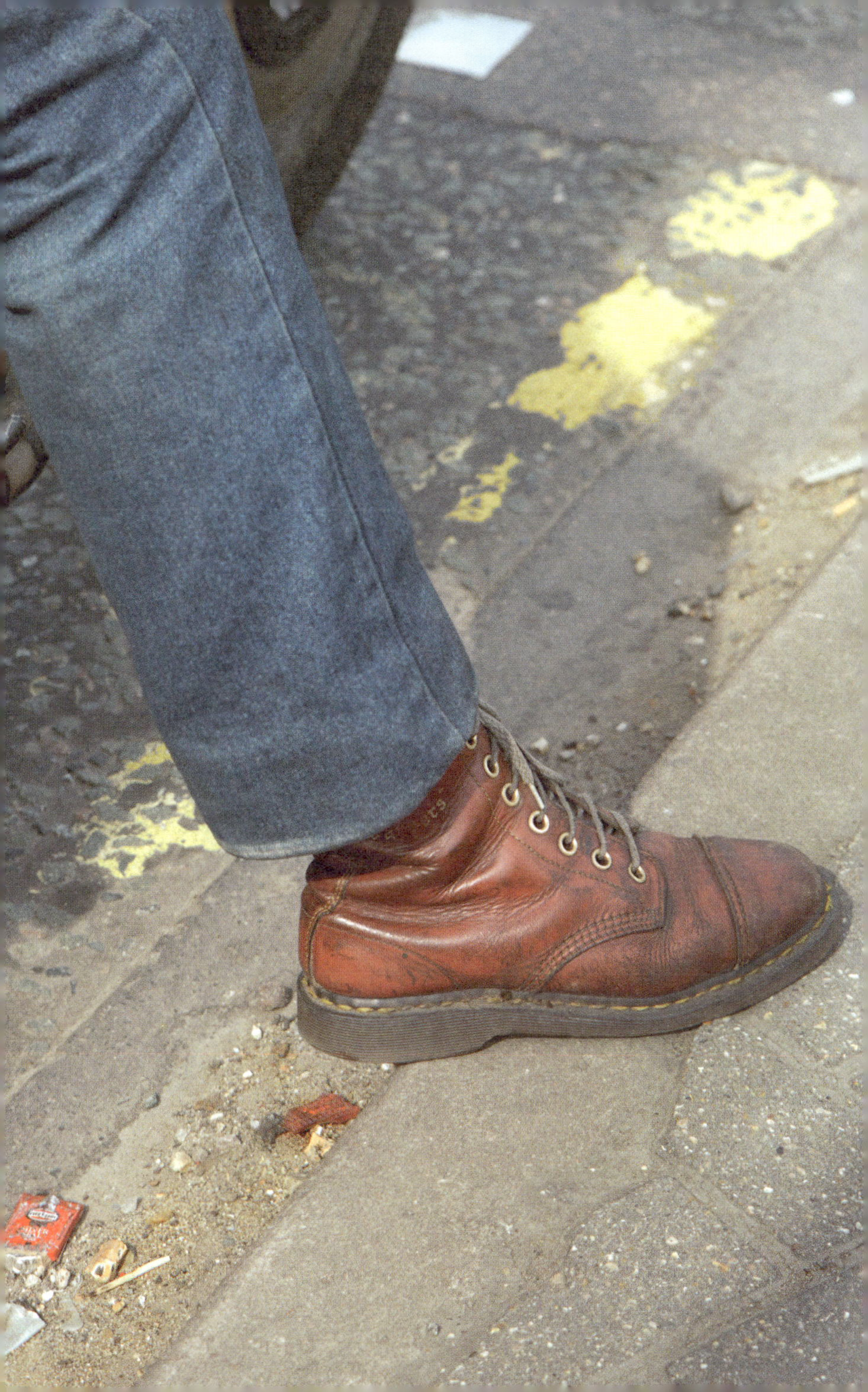

Austin Maestro, London Wall, EC2

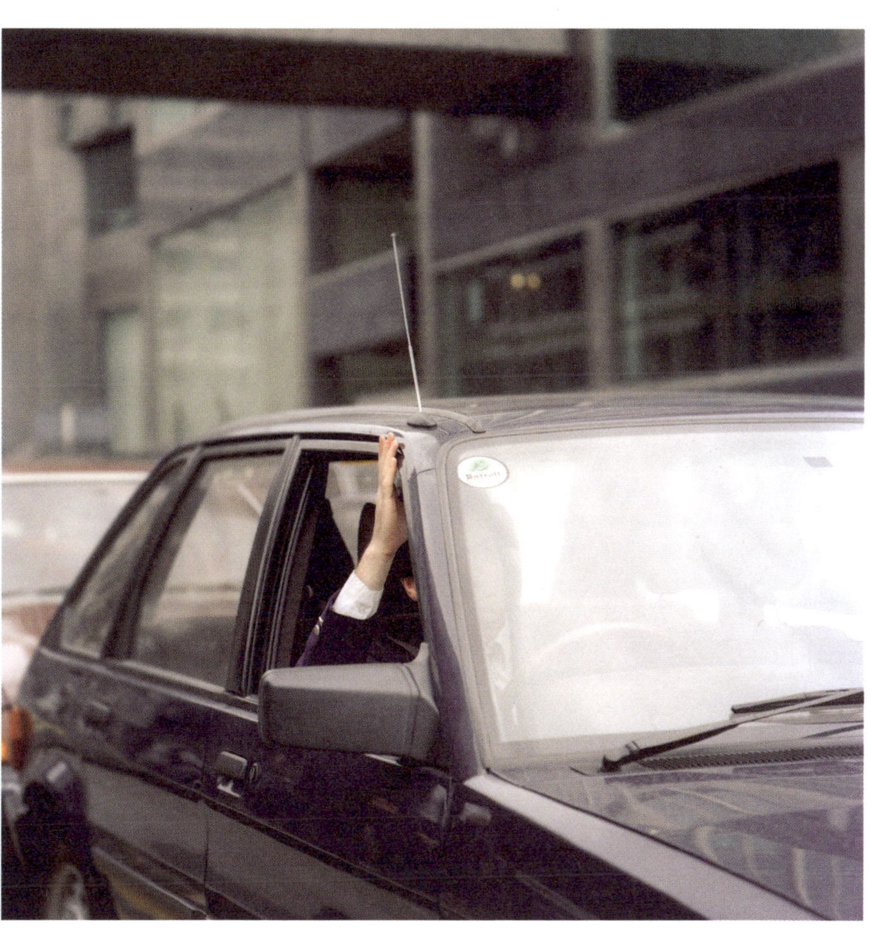

Mercedes-Benz W115 Series, London Bridge, EC4

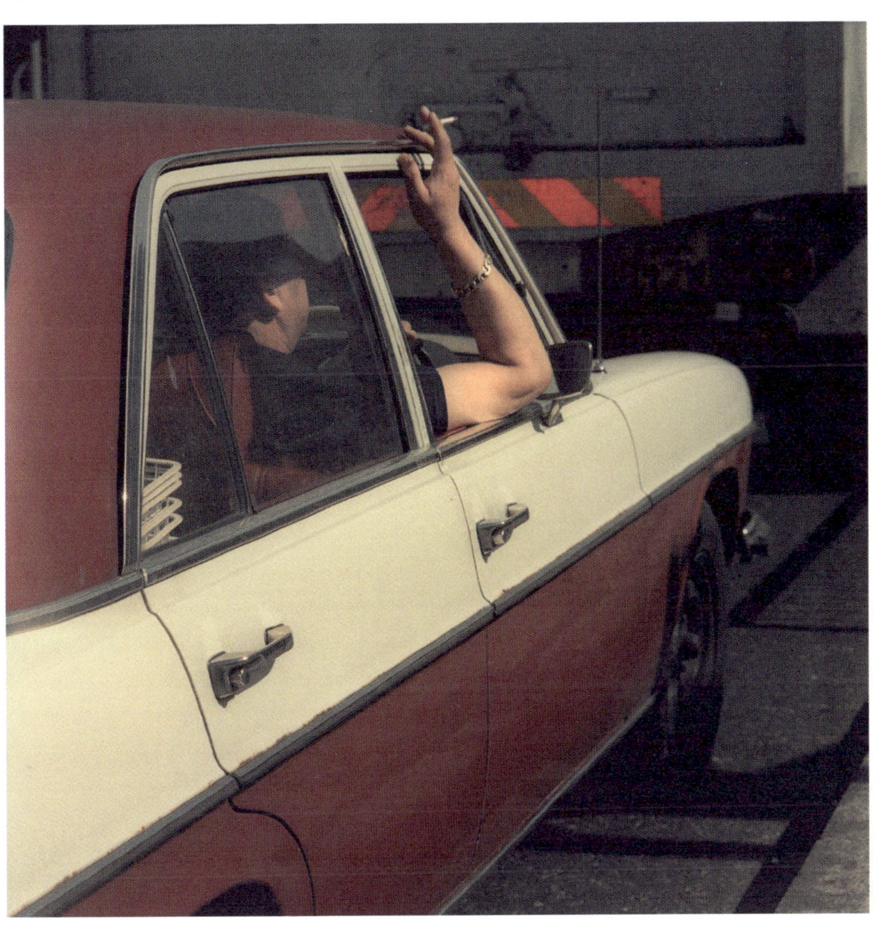

Ford Sierra XR4i, Bishopsgate, EC2

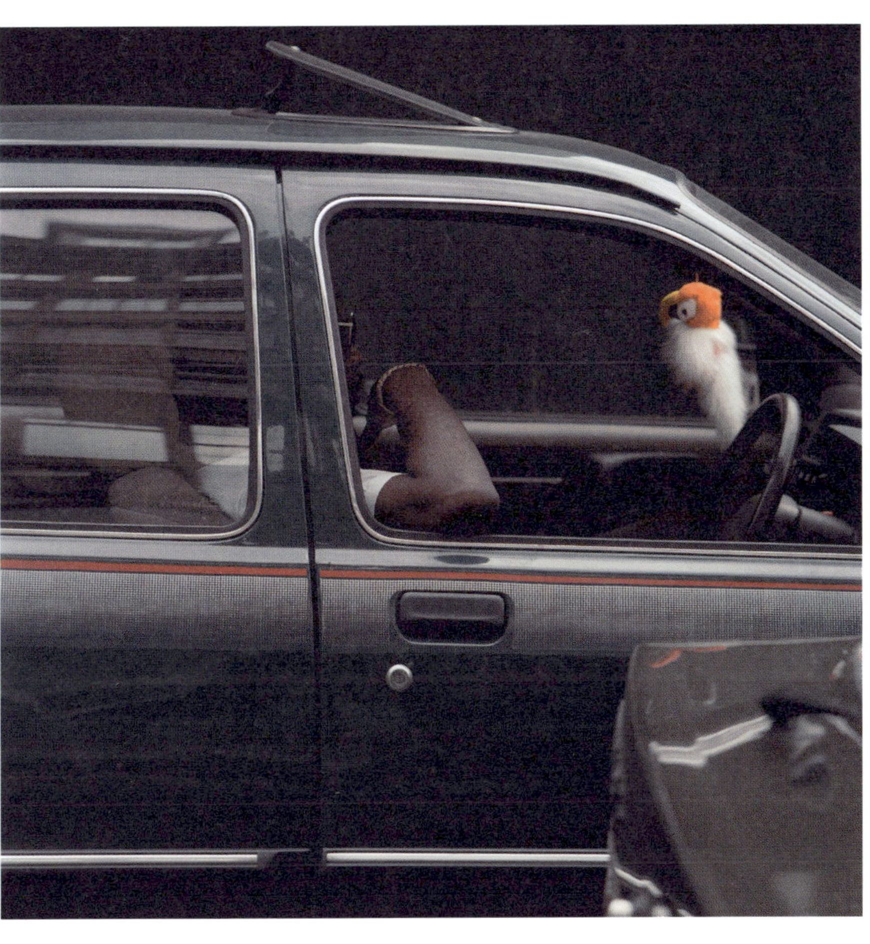

Datsun 120Y, Old Street, EC2

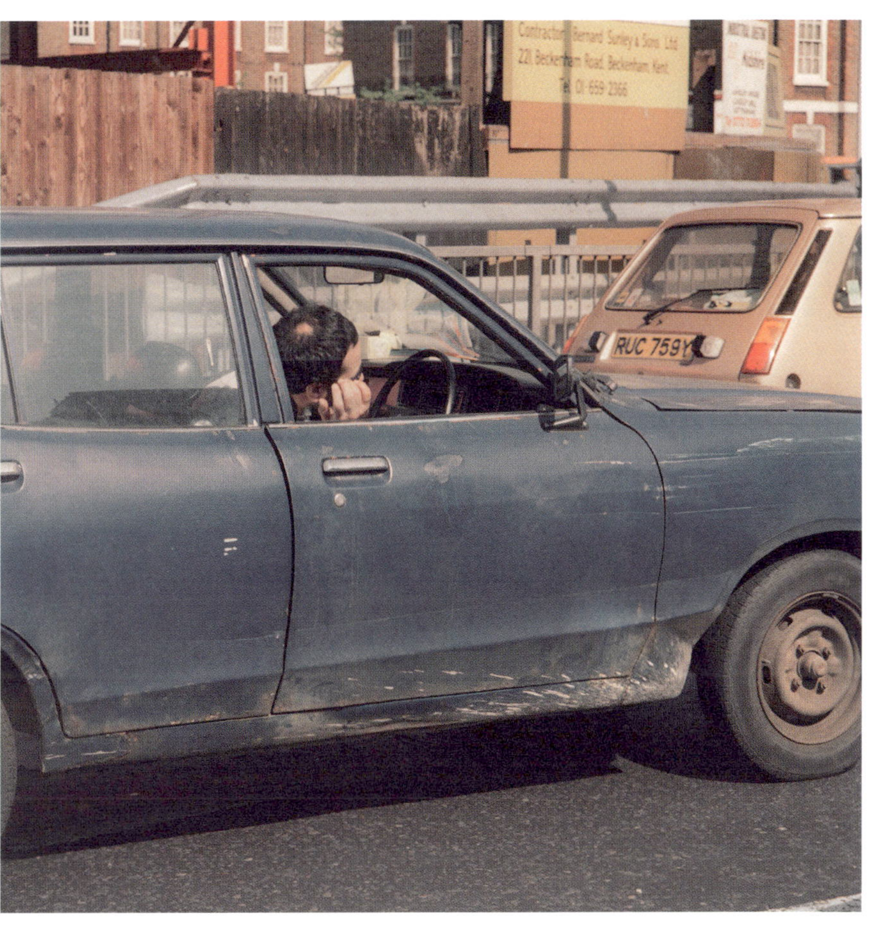
Bernard Sunley & Sons Ltd
Road, Beckenham, Kent.
01-659-2366
RUC 759Y

Toyota Corolla AE82, London Bridge, EC4

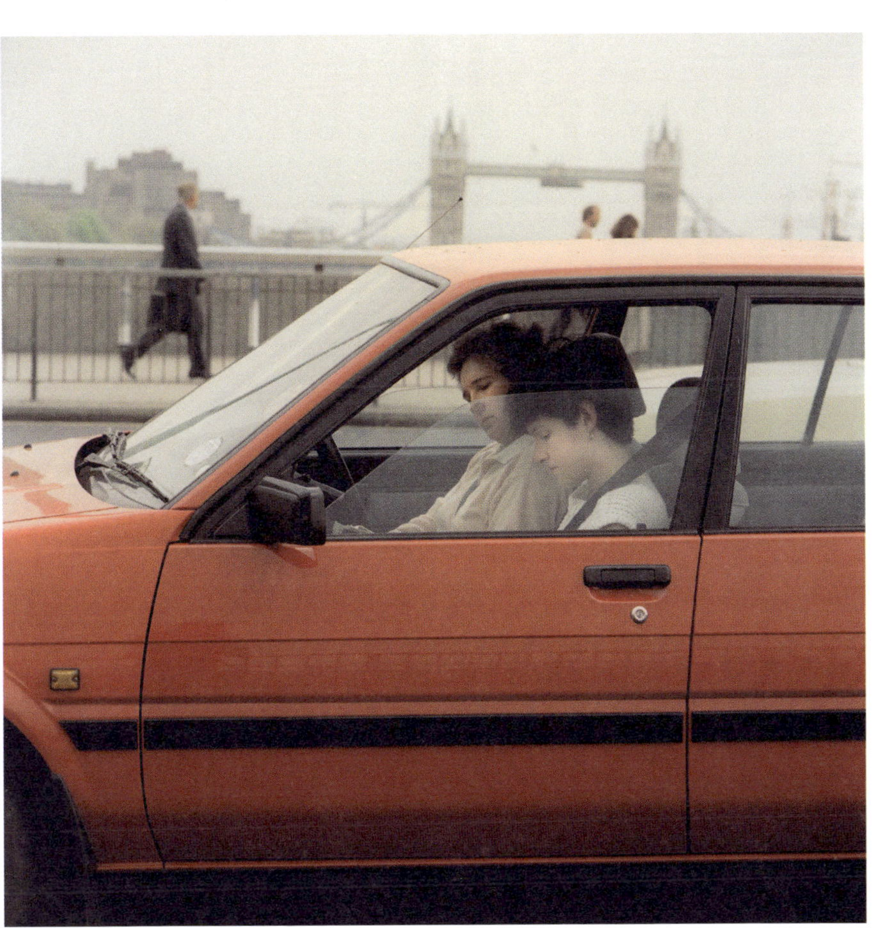

Black Cab LTI FX4, Bank, EC2

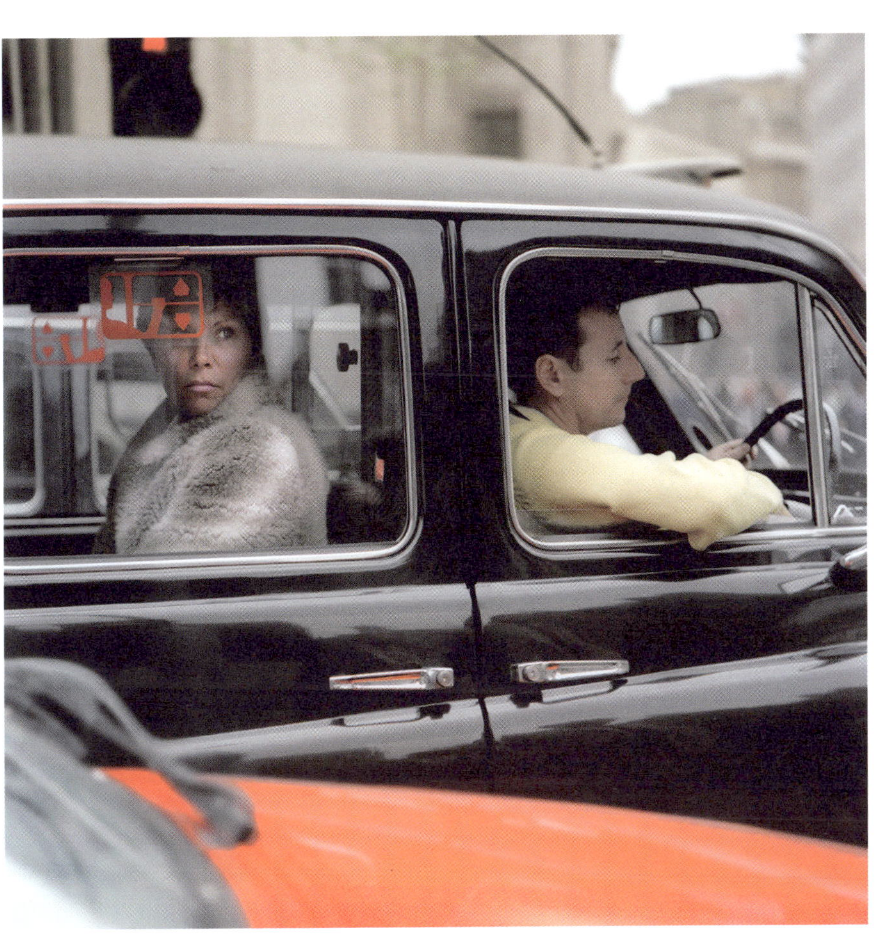

Ford Escort Mk3 Van, London Wall, EC3

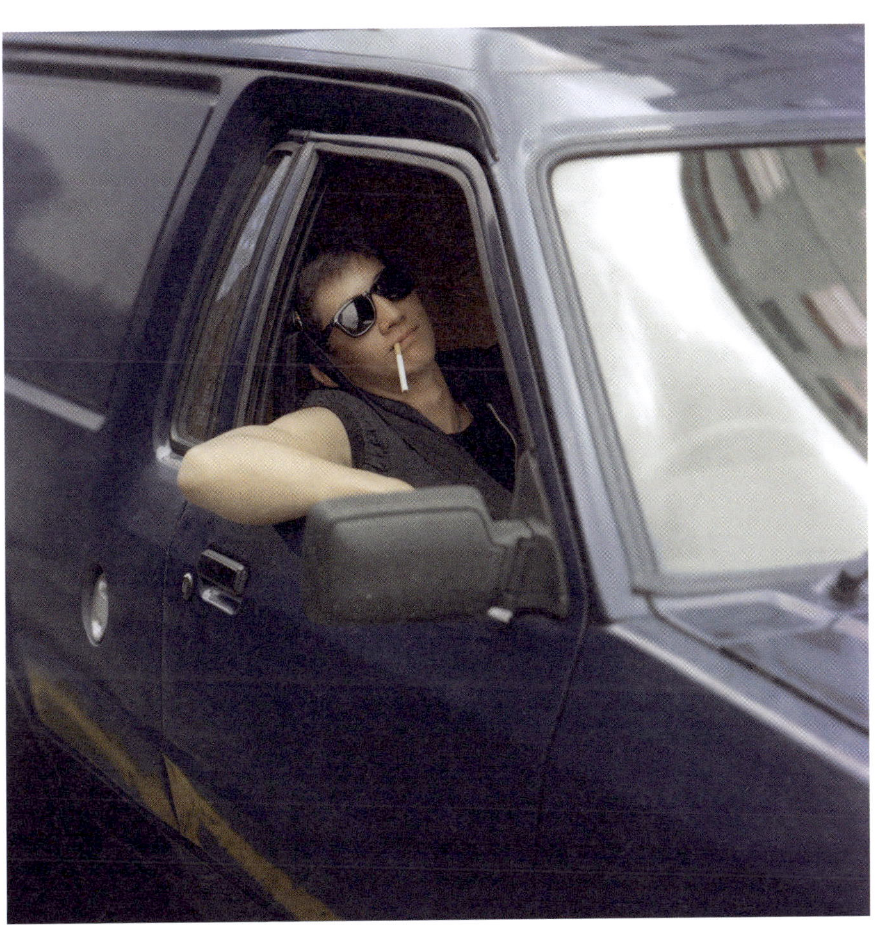

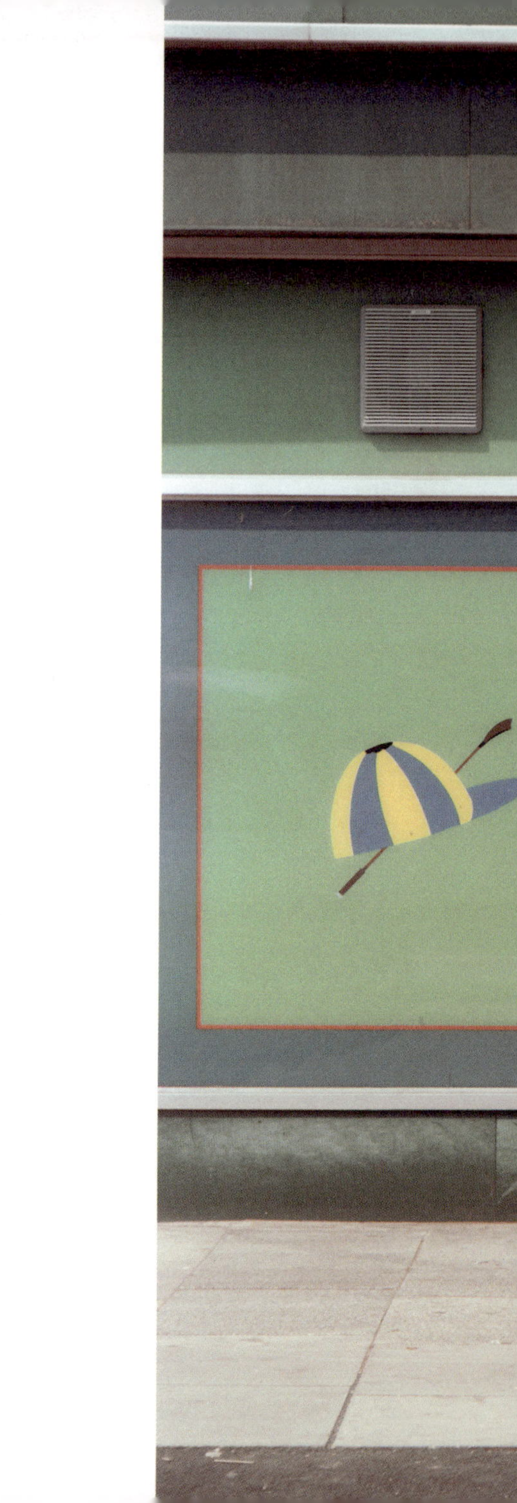

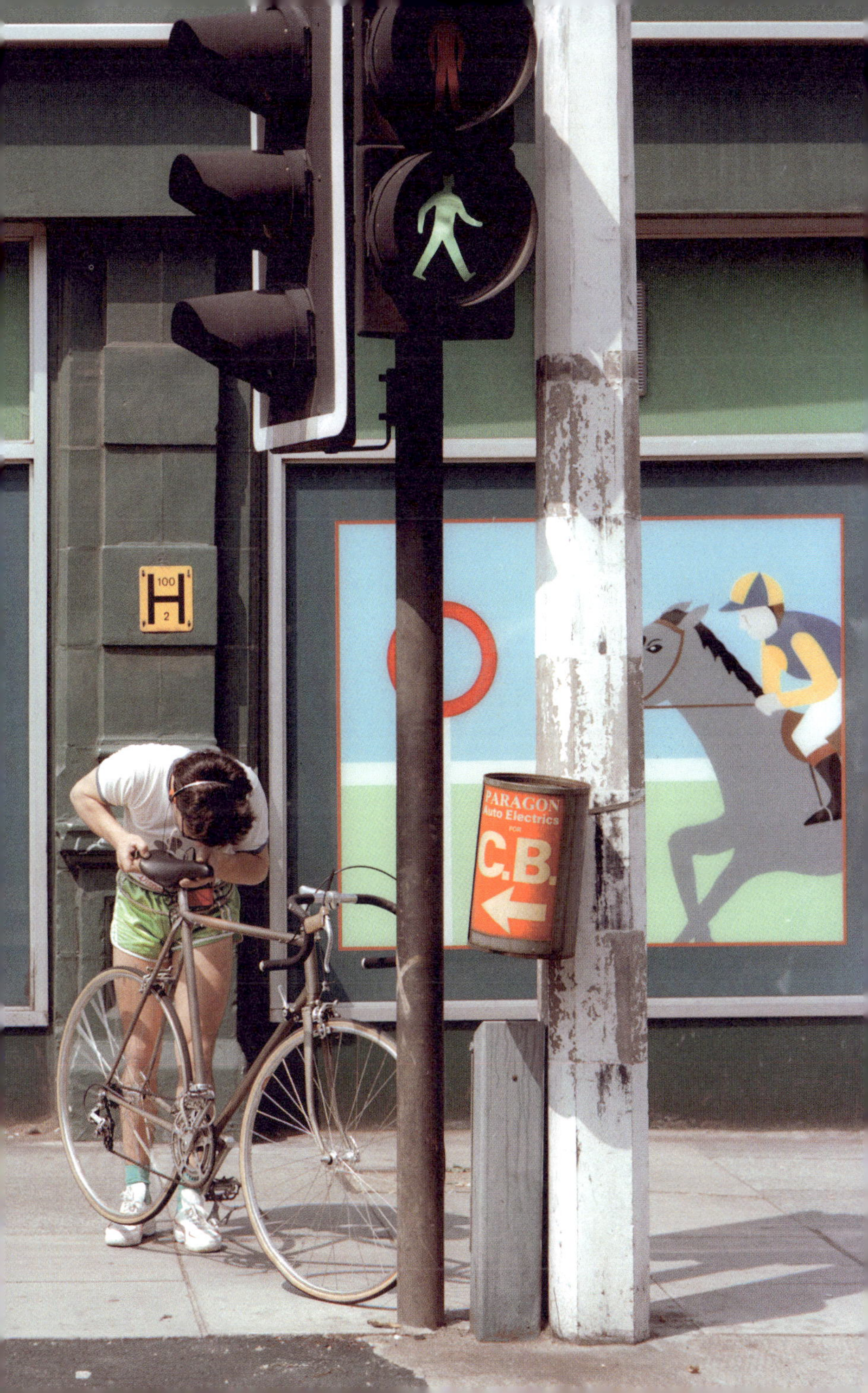
100
H
2
PARAGON
Auto Electrics
FOR
C.B.

Renault 30, Mare Street, E8

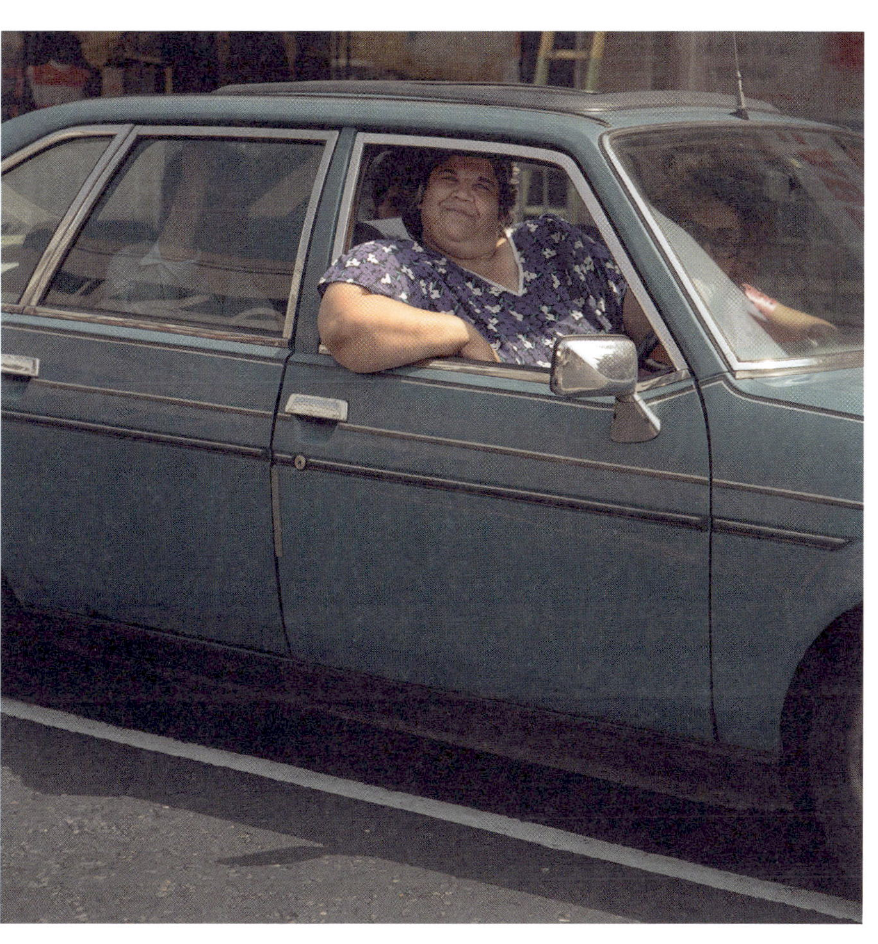

Cortina Mk5 Crusader, Mare Street, E8

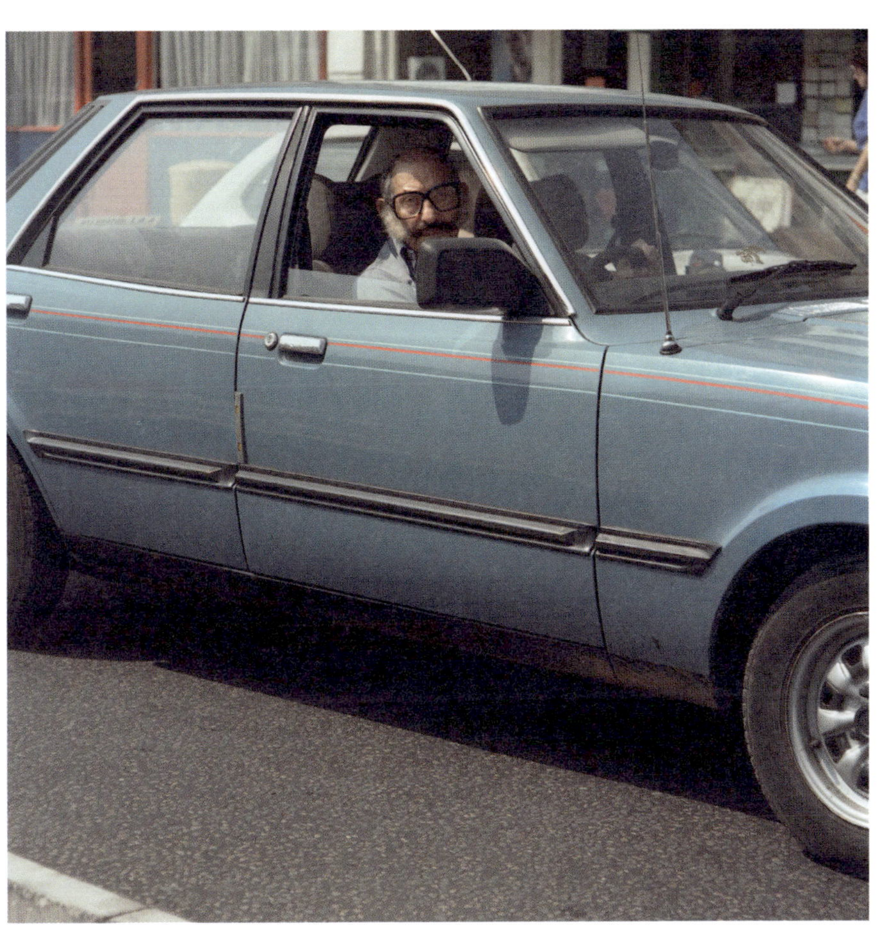

Triumph Spitfire, Mare Street, E8

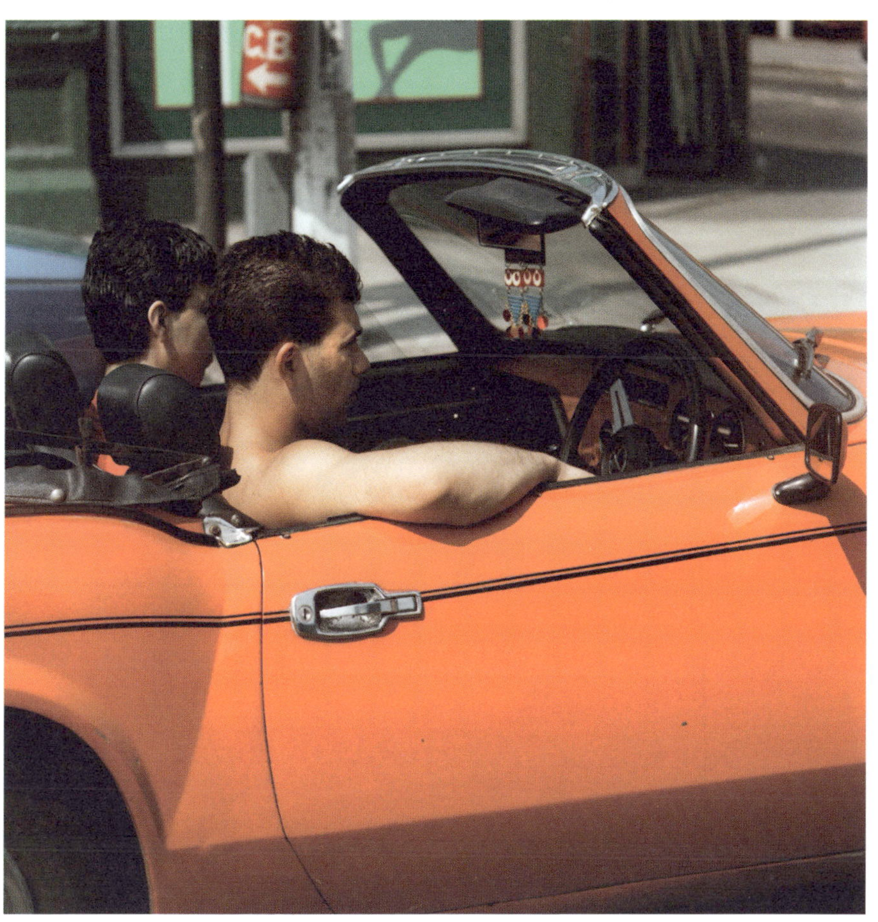
C.B

Bedford Plaxton Coach, London Bridge, EC4

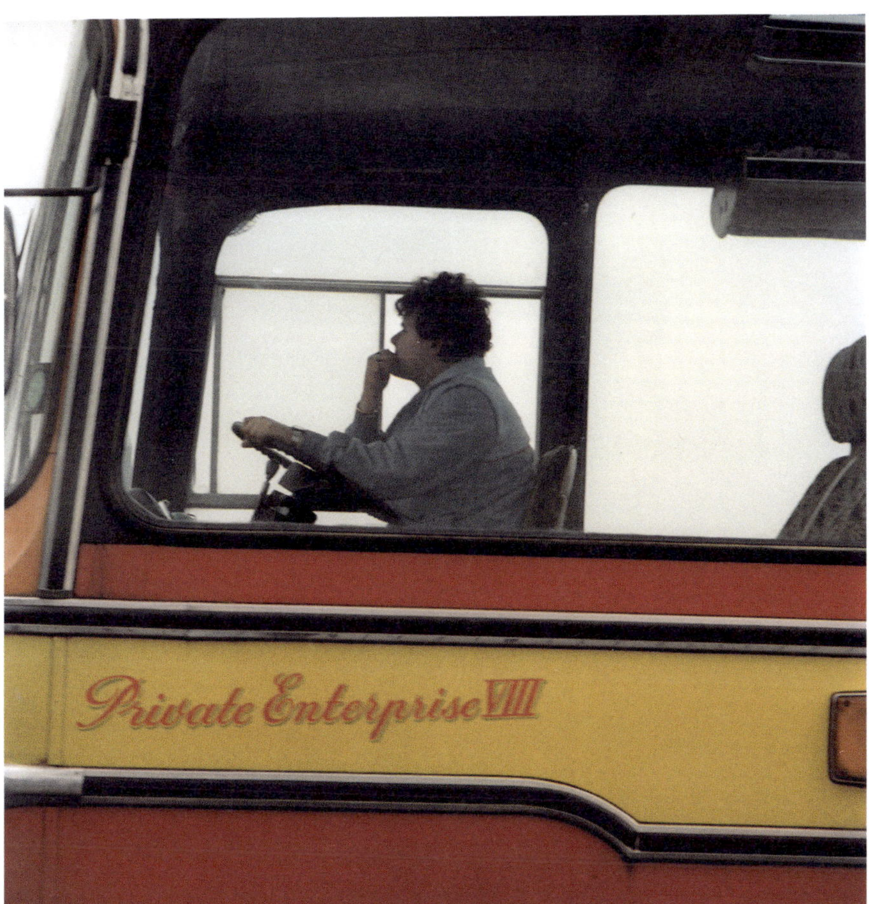
Private Enterprise VIII

Willys Jeep, Hackney Road, E2

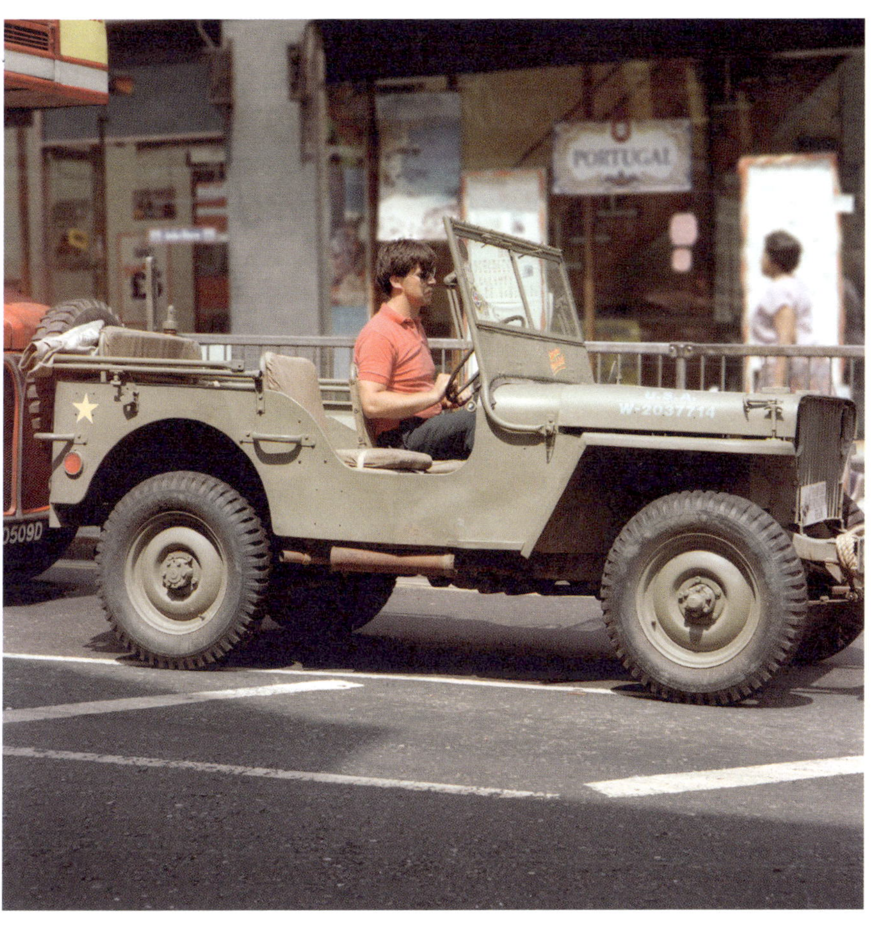
PORTUGAL
U.S.A.

Peugeot 305, Old Street, EC2

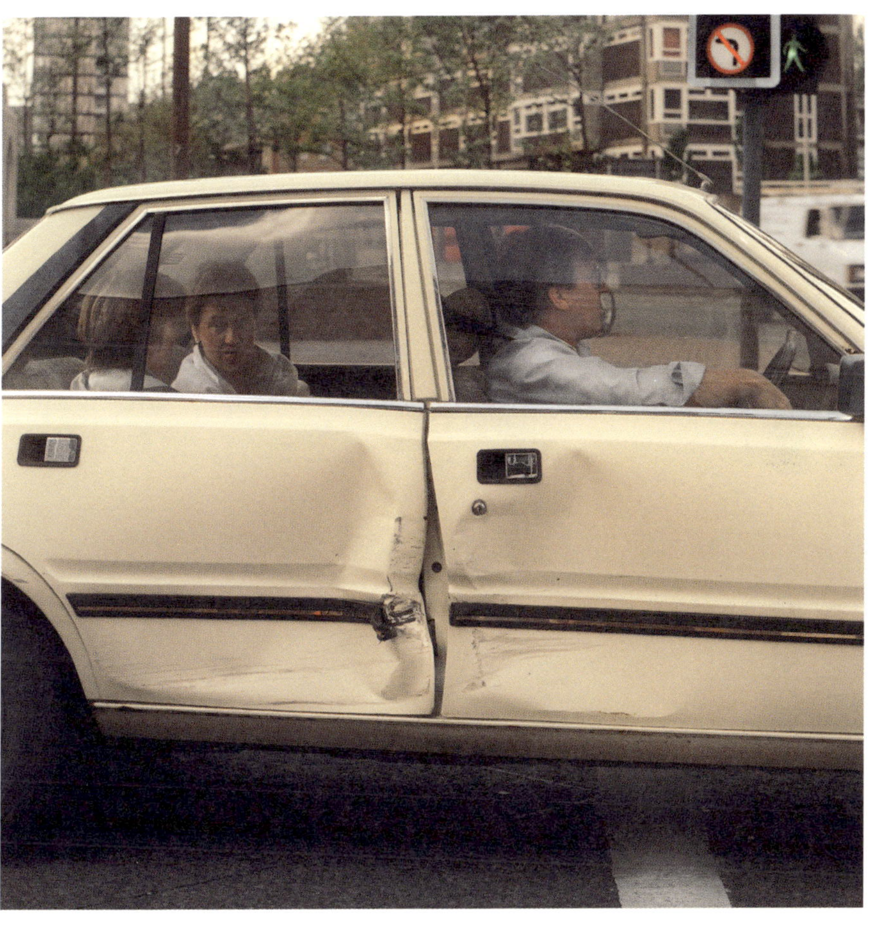

Ford Sierra, London Bridge, EC4

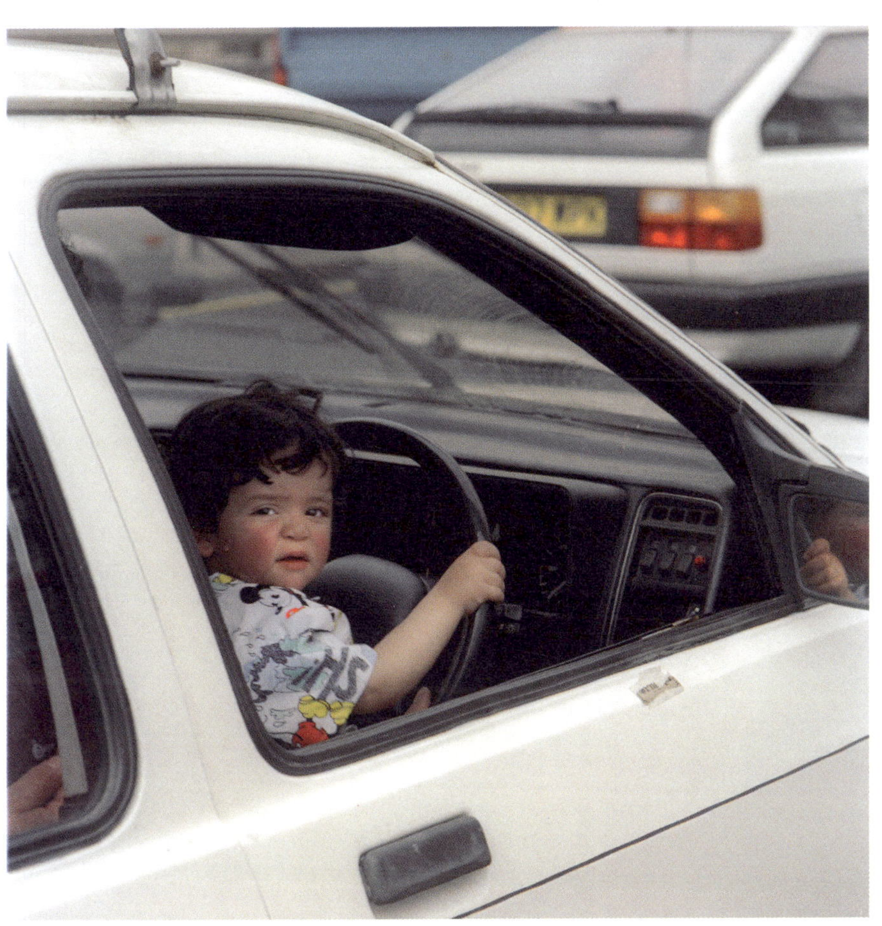

Ford Capri, London Bridge, EC4

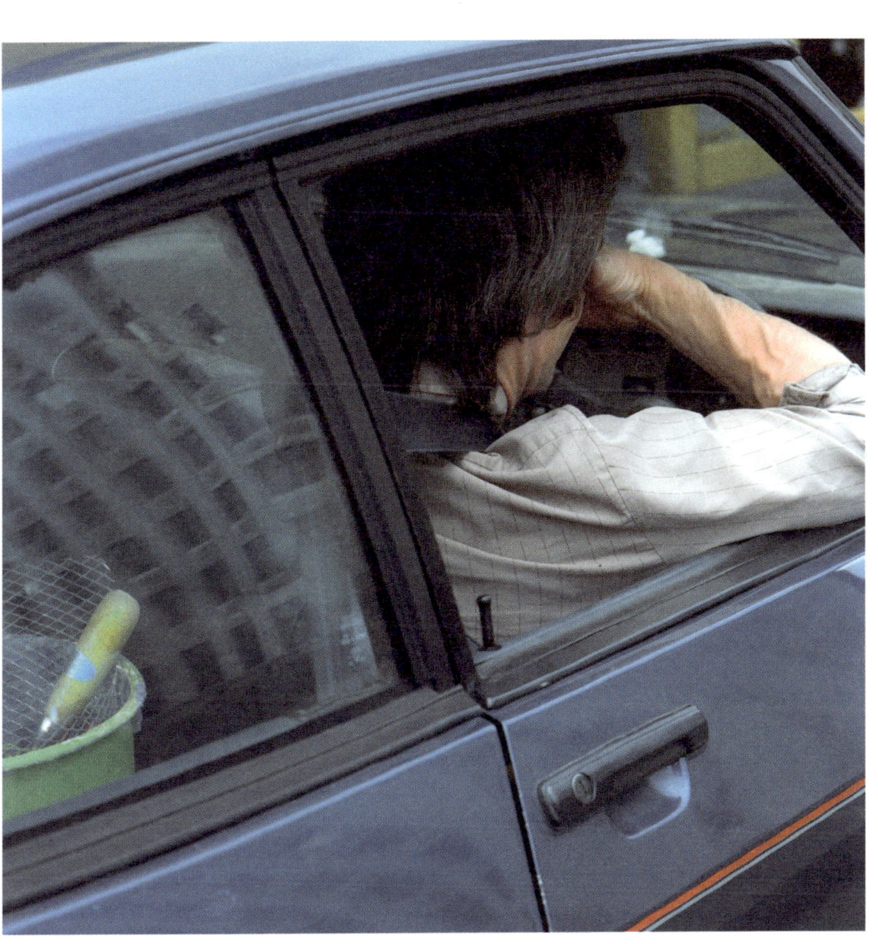

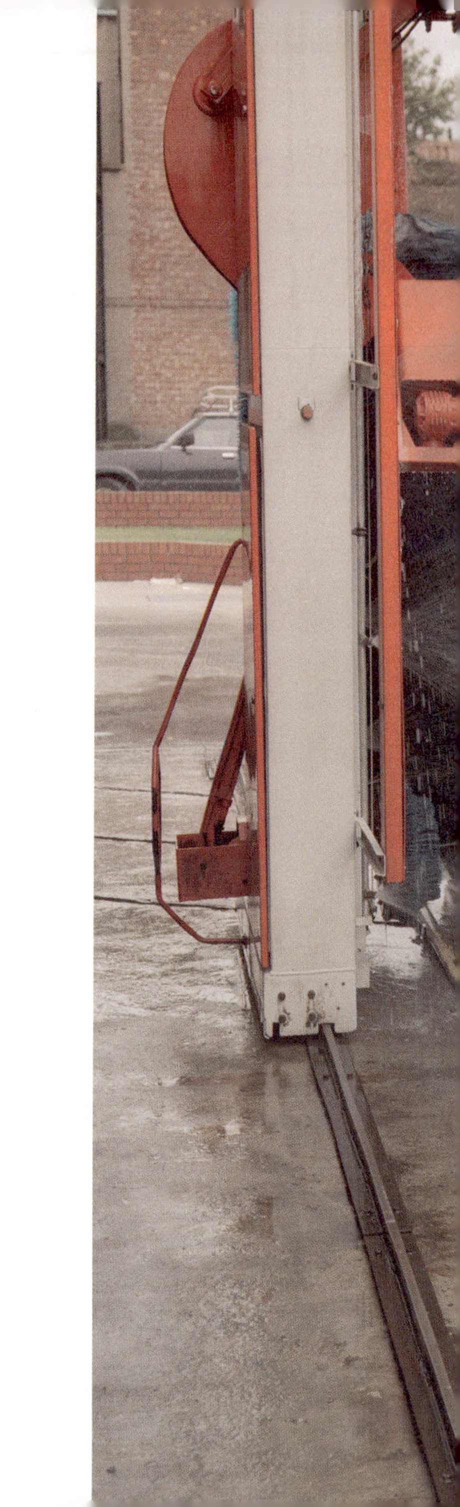

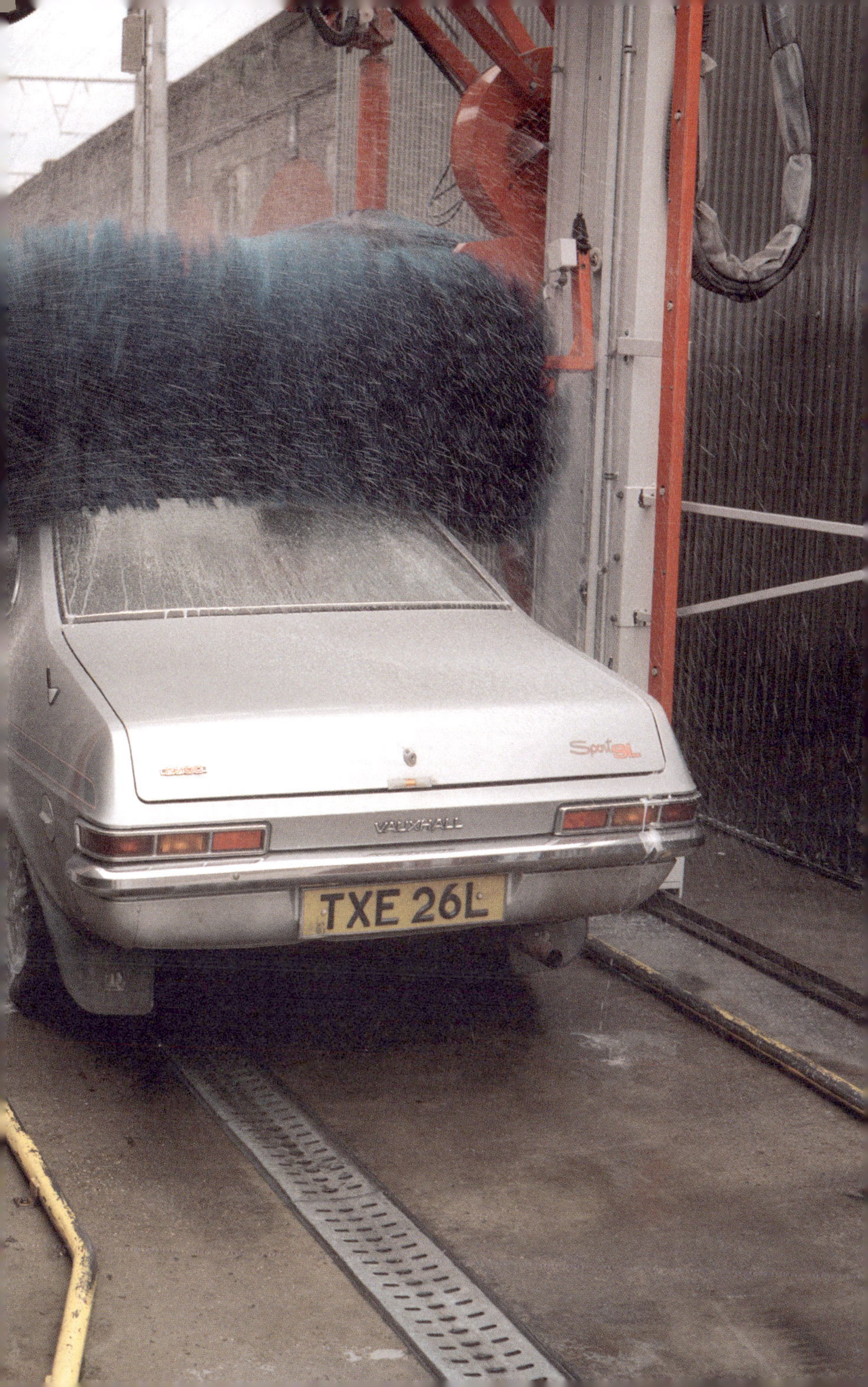
VAUXHALL
TXE 26L

Drivers in the 1980s
First Edition

All photographs © Chris Dorley-Brown
Design and sequence by Martin Usborne and Chris Dorley-Brown
Series design by breadcollective.co.uk

A CIP catalogue record for this book is available from the British Library.

ISBN: 978-0-9576998-9-2

First published in the United Kingdom in 2015 by Hoxton Mini Press

Printed and bound by: RRD, China

To order books, collector's editions and signed prints please go to:
www.hoxtonminipress.com